U0077524

博碩文化

博碩文化

博碩文化

覺得 Kubernetes 門檻太高？那你找對地方了！

從異世界歸來
發現只剩自己不會
Kubernetes

初心者進入雲端世界
的實戰攻略！

許宏翔（Mike Hsu）著

從敬而遠之到心領神會！

九大核心主題，由淺入深逐一擊破

漸進式學習	主題式攻略	提供實作範例	過來人甘苦談
艱澀觀念都 能迎刃而解	專題深度剖析 策略全面掌握	大量實作範例 讓你寫得順看得懂	記錄從學廢到 學會的心路歷程

2022 iThome 鐵人賽 優選

iThome 鐵人賽

本書如有破損或裝訂錯誤，請寄回本公司更換

作　　者：許宏翔（Mike Hsu）
責任編輯：Lucy

董 事 長：陳來勝
總 編 輯：陳錦輝

出　　版：博碩文化股份有限公司
地　　址：221 新北市汐止區新台五路一段 112 號 10 樓 A 棟
　　　　　電話 (02) 2696-2869　傳真 (02) 2696-2867

發　　行：博碩文化股份有限公司
郵撥帳號：17484299　戶名：博碩文化股份有限公司
博碩網站：http://www.drmaster.com.tw
讀者服務信箱：dr26962869@gmail.com
訂購服務專線：(02) 2696-2869 分機 238、519
（週一至週五 09:30 ～ 12:00；13:30 ～ 17:00）

版　　次：2024 年 2 月初版二刷

建議零售價：新台幣 650 元
I S B N：978-626-333-749-7
律師顧問：鳴權法律事務所 陳曉鳴律師

國家圖書館出版品預行編目資料

從異世界歸來發現只剩自己不會Kubernetes
：初心者進入雲端世界的實戰攻略！/ 許宏翔
(Mike Hsu)著. -- 初版. -- 新北市：博碩文化股
份有限公司, 2024.02印刷
　　面；　公分. -- (iThome鐵人賽系列書)

ISBN 978-626-333-749-7 (平裝)

1.CST: 作業系統 2.CST: 軟體研發

312.54　　　　　　　　　　　　113001121

Printed in Taiwan

博碩粉絲團　歡迎團體訂購，另有優惠，請洽服務專線
　　　　　　(02) 2696-2869 分機 238、519

商標聲明

本書中所引用之商標、產品名稱分屬各公司所有，本書引用
純屬介紹之用，並無任何侵害之意。

有限擔保責任聲明

雖然作者與出版社已全力編輯與製作本書，唯不擔保本書及
其所附媒體無任何瑕疵；亦不為使用本書而引起之衍生利益
損失或意外損毀之損失擔保責任。即使本公司先前已被告知
前述損毀之發生。本公司依本書所負之責任，僅限於台端對
本書所付之實際價款。

著作權聲明

本書著作權為作者所有，並受國際著作權法保護，未經授權
任意拷貝、引用、翻印，均屬違法。

推薦序

身為一名後端工程師，我經常被問到：「在日常的後端開發工作中，為何需要學習 Kubernetes ？」對我來說，答案是顯而易見的。在這個微服務和容器化部署成為標準的時代，掌握 Kubernetes 不僅能讓我們的應用更加穩定、可擴展，還能大大提高開發和部署的效率。這正是我極力推薦 Mike 寫的這本《從異世界歸來發現只剩自己不會 Kubernetes》的原因。

後端工程師的工作內容不只是開發程式，更重要的是確保應用在任何環境中都能夠順暢運行。而 Kubernetes 提供了這樣的平台，讓我們在開發、測試和生產環境中，都能確保應用的高可用性和可擴展性。從這個角度來看，Kubernetes 不僅僅是一個維運工具，更是後端開發者的得力助手。

當我首次接觸 Kubernetes 時，面對其龐大的生態系和複雜的設定，確實感到有些手足無措。但是，當我開始翻閱這本書，所有的困惑和疑惑都一一有了答案。作者從後端開發的視角出發，深入淺出地解釋了 Kubernetes 的核心概念、工作原理以及與後端開發的緊密聯繫。

這本書不僅提供了豐富的範例程式碼和操作指南，讓身為工程師的我們能實際操作來加深認知；更重要的是，它教會我如何從後端工程師的角度去思考和應用 Kubernetes。從容器的生命週期、資源管理到部署管理，每一章都與我們的日常開發工作息息相關。

身為後端工程師，我們經常需要考慮到應用的效能、可靠性和安全性。而 Kubernetes 提供了一套完整的解決方案，幫助我們輕鬆應對這些挑戰。透過這本書，我學會了如何利用 Kubernetes 的強大功能，打造出高效、穩定且易於維護的後端系統。

對於還在猶豫是否要學習 Kubernetes 的後端工程師，我想說：這是一本值得你
投入時間的書籍。它將會帶領你進入 Kubernetes 的異世界，更會幫助你進步成
一名更優秀的後端開發者。與其在疑惑和困難中浪費時間，不如拿起這本書，
開始你的學習之旅吧！

雷 N

後端工程師 / iThome 鐵人賽戰友

序

關於異世界歸來的起點

這是一個異世界歸來後發現只剩自己不會 Kubernetes 的故事，一切都要從非相關科系的我踏入網頁前後端，並且在愚昧山丘待了一年多開始說起。

一開始對於小白時期的我，寫寫 API 串串資料讓畫面可以互動起來，就已經滿足我對網路世界的想像，那時候對於容器化的概念，也只停留在工作上已經有現成的環境，會用、會啟動就好。對於其他可能的應用場景沒有過多了解。直到接觸了 Golang 與 Node 這等適合拿來做微服務的語言，接觸到微服務概念的我彷彿就像大夢初醒一樣，瞬間將我從異世界拉回現實。它是一種軟體架構風格，以專注於特定功能切分成各種服務，並且以組合的方式建構出複雜的大型應用。

為了管理如此龐大複雜的應用，需要輕便、彈性、有效率的容器化技術來管理這些服務，這時候 Kubernetes 在 2015 年橫空出世帶來爆炸式的創新，提供了一個可以共同部署維護並擴充的機制，不但有效的降低耦合，並且可延伸滿足不同的工作負載，雖然我不是數學家，但這真的深深地令我著迷！

掌握未來：深化對 Kubernetes 在容器化時代的理解

在我踏入資訊技術的世界以來，我始終在尋找那些能夠讓我們更好地理解和利用這個領域的關鍵概念。這就是我寫這本書的一個主要動機。我認為 Kubernetes 是當代技術環境中一個重要且不能忽視的元素，特別是在我們進入所謂的「容器化時代」時。

對於 Kubernetes，有許多現成的資源和教材，但我發現許多材料都偏向於在理論上的解釋，而對實際操作的指導和理解卻相對缺乏。我的觀點是，理論和實際應用都是不可或缺的，尤其是在如此實用且多變的領域，應該更注重在如何降低絕大多數人踏入這個 Kubernetes 的門檻。

透過這本書，我希望能夠將 Kubernetes 的理論與實踐結合起來，並以我個人的角度，從基礎概念開始介紹，逐步深入到更複雜的主題。我想讓讀者不僅了解 Kubernetes 是什麼，更重要的是，我希望他們能夠了解 Kubernetes 的重要性，以及如何在實際的工作或項目中使用它。我相信這種理論與實踐相結合的方式，將使讀者能夠更全面地理解 Kubernetes，並能夠在容器化時代中立足。

成為 Kubernetes 高手：本書將讓你獲得什麼

本書的目標是帶領你進入 Kubernetes 的世界，從基本概念到實際操作，進階主題，甚至是專題研究，都在本書的範疇之內。

首先，你將從本書「概念篇」和「安裝篇」中了解 Kubernetes 的基礎，瞭解它是什麼以及如何安裝。接著在「基礎篇」和「進階基礎篇」中，我們將踏入實際的應用操作，學習如何使用 Kubernetes 的基本元件，例如 Pod、Service、Deployment，並且了解如何使用路由與其他功能。

然後，我們將在「實戰部署篇」中學習如何透過不同的部署策略來實現應用的順利運行。接下來的三個篇章 —「Volume」、「Resources」和「AutoScaling」，則是對 Kubernetes 的進階主題進行深入的探討，涵蓋資源監控、自動縮放、資源分配等多個主題。

最後，在「Security」篇章中，我們將討論 Kubernetes 的安全性，學習如何管理用戶並進行權限管理，使你對 Kubernetes 的安全有更全面的理解。

在這一路的學習過程中,我相信你將得到深刻的理解,並得到實際的技能,讓你在面對 Kubernetes 的挑戰時,能夠信心滿滿。此外,我也希望你能透過本書,看到 Kubernetes 的可能性和潛力,並開始思考如何在你自己的項目或工作中適應和利用這些技能。

最後,我期待你在閱讀完這本書之後,能帶著你的新知識和新技能,走向 Kubernetes 的旅程,並在這個快速變化的資訊世界中,找到你自己的定位。我期待見證你的成長,並期待你將來能夠分享你的經驗和成功,使我們的 Kubernetes 社群更加豐富多元。

免責聲明

本書的內容基於網路公開資訊及官方文件,儘管我們已竭力確保準確性,但無法保證所有資訊始終完全正確或適用。本書僅供參考,使用其資訊和建議的風險由讀者自行承擔。如遇到問題,請尋求專業意見或查詢官方文件。作者與出版商對於任何損失概不負責。

本書的閱讀方法

本書旨在帶領讀者體驗 Kubernetes 的實作過程，我們將一起探索其結構、功能以及豐富的用途。在每一章節的結尾，我都會提供對應的程式碼以供各位參考，而這些程式碼都已整理在我的 GitHub 專案中。

為了讓大家能更順利地找到相關資源，請先到我的 GitHub 專案：kubernetes-from-another-world 中。在該專案中，你將看到一系列與本書章節對應的資料夾，每個資料夾中都包含了該章節所需的全部程式碼。

此外，書中也會包含一些相關的網址，我將這些網址轉換成 QR Code，只需掃描一下，就能直接打開相關資源。

GitHub 專案連結：

https://github.com/MikeHsu0618/kubernetes-from-another-world

▲ GitHub Repository

透過這種方式，我希望能讓閱讀本書的過程成為一次實際與理論相輔相成的學習旅程，讓你更深入地理解 Kubernetes 的運作原理與其強大之處。

礙於本書篇幅有限，書中的範例檔僅展示部分程式碼，欲查看完整的程式碼，請前往上方的 GitHub 連結或掃描 QR Code。

適合讀者

這本書提供多種需求，滿足初學者對於 Kubernetes 的各種需求。

- **適合初學者**：本書針對初學小白設計，內容深入淺出，讓讀者能快速掌握 Kubernetes 的基礎概念和實踐技巧。

- **範例豐富**：本書不僅提供了豐富的理論知識，還提供了大量的實例，讓讀者可以輕鬆地進行實踐。

- **知識體系全面**：本書涵蓋了 Kubernetes 的主要元件、架構、原理、操作、除錯等各個方面的知識，讓讀者可以全面深入地了解 Kubernetes。

- **最新技術**：本書關注最新的 Kubernetes 技術，並介紹了 Kubernetes 的最新特性和趨勢。

- **實戰為主**：本書注重實戰，針對實際場景提供了解決方案，讓讀者能夠快速掌握 Kubernetes 的實戰技能。

- **可擴展性強**：本書涵蓋了 Kubernetes 的擴展性，讓讀者可以掌握如何在 Kubernetes 上部署不同種類的應用程式和服務。

- **深入細節**：本書不僅關注 Kubernetes 的概念，還深入到 Kubernetes 的實現細節，讓讀者了解 Kubernetes 的核心技術和實現方式。

- **應用場景廣泛**：本書不僅介紹了 Kubernetes 的基礎概念和實踐技巧，還針對不同的應用場景提供了解決方案，讓讀者能夠更好地應用 Kubernetes。

本書將目標讀者鎖定在以下族群，相信能更多的讓各位了解本書是否合適各位
讀者：

- **後端工程師**：已有 Docker 或微服務基礎，想更深入了解 Kubernetes。

- **DevOps 工程師**：想要在 Kubernetes 上部署和管理應用程式的 DevOps 工
  程師，可以透過本書快速掌握 Kubernetes 的基礎概念和實踐技巧，從而更
  好地完成工作。

- **軟體工程師**：想要學習 Kubernetes 的軟體工程師，可以透過本書了解
  Kubernetes 的原理和架構，以及如何在 Kubernetes 上部署和運行應用程
  式。

- **系統管理員**：想要學習 Kubernetes 的系統管理員，可以透過本書深入理解
  Kubernetes 的各個元件和操作方式，從而更好地管理 Kubernetes 環境。

- **技術管理者**：想要深入了解 Kubernetes 技術，以便能更加掌握如何在企業
  中應用 Kubernetes 的技術管理者，可以通過本書瞭解 Kubernetes 的基礎知
  識和實踐技巧，並了解如何在企業中推廣和應用 Kubernetes。

- **對 Kubernetes 感興趣的 IT 從業者**：對 Kubernetes 感興趣的 IT 從業者，
  可以透過本書了解 Kubernetes 的基礎知識和實踐技巧，從而更好地掌握這
  一熱門技術。

你期望完成的目標

你是否對網路世界的博大精深充滿了好奇，就像我對 Kubernetes 異世界深深的著迷？或者你有一項具體的目標，期待透過深入研究這本書來達成？無論你的動機是什麼，設定一個明確的目標可以大大提升你的學習效率和動力。

我一直都是一個熱愛學習和閱讀的人。但坦白說，我發現很多書籍購買回家後，往往只會成為書架上的裝飾品，證明我們的知識淵博，卻未必實質增益我們的學問。因此，在你開始閱讀這本書之前，我希望你能問問自己：「我有什麼具體的目標或任務，希望透過學習 Kubernetes 來實現？」這可能是完成一個創新的專案，或者是對網路世界更深入的探索。請在下方將你的目標寫出：

_____

把你的想法和目標化為文字，每次翻開這本書時，它就會提醒你，你正為了一項重要的任務而學習，提升你的學習動力和效果。網路世界的博大精深，以及 Kubernetes，都在等著我們去探索和掌握。

目錄

Part 1 概念篇：萬丈高樓平地起，開始爬吧！

01 Kubernetes 是什麼？

1.1　網路部署的演變 ……………………………………… 1-2

1.2　Kubernetes 的定位 …………………………………… 1-4

02 Kubernetes 的元件

2.1　Kubernetes 設計原理 ………………………………… 2-2

2.2　Kubernetes 叢集中包含哪些元件？ ………………… 2-2

2.3　Kubernetes Control Plane …………………………… 2-4

Part 2 安裝篇：一定要安裝些什麼的吧！

03 安裝 Kubernetes（Docker Desktop）

3.1　下載 Docker Desktop（macOS）……………………… 3-2

3.2　開啟 Docker Desktop 內建的 Kubernetes …………… 3-4

3.3　查看 Kubernetes 狀態 ………………………………… 3-5

04 安裝 Kubernetes Dashboard GUI

4.1　Kubernetes Dashboard 是什麼？ …………………… 4-2

4.2　設定 Kubernetes Dashboard ………………………… 4-3

Part 3 基礎篇：老闆總說先可以 Run 就好…

05 Kubernetes — 實戰做一個 Pod

5.1　建立容器（Container）………………………………… 5-2

5.2 建立 Kubernetes 設定檔 ……………………………………… 5-5

5.3 在 Kubernetes 中建立 Pod ……………………………………… 5-6

5.4 使用 kubectl port-forward 與 Local 端接軌 …………………… 5-8

06 Kubernetes — 實戰做一個 Service

6.1 Service 是什麼？ ……………………………………………… 6-2

6.2 那什麼是邏輯上的一群 Pod？ ………………………………… 6-2

07 Kubernetes — 實戰做一個 Deployment

7.1 使用案例 ………………………………………………………… 7-3

7.2 實戰演練 ………………………………………………………… 7-3

7.3 更新 Deployment 實現水平擴展 ……………………………… 7-6

7.4 使用 Rollout 查看歷史版本並回滾 …………………………… 7-9

08 Kubernetes — 實戰做一個 StatefulSet

8.1 StatefulSet 是什麼？ …………………………………………… 8-2

8.2 StatefulSet 中的有序命名及網路 ID …………………………… 8-3

8.3 StatefulSet 中的穩定儲存 ……………………………………… 8-4

8.4 StatefulSet 中的 Headless Services …………………………… 8-5

8.5 StatefulSet 中的部署及擴縮保證 ……………………………… 8-6

8.6 StatefulSet 中的更新策略 ……………………………………… 8-7

8.7 實戰演練 ………………………………………………………… 8-8

8.8 刪除 StatefulSet ………………………………………………… 8-15

Part 4 進階基礎篇：我就知道事情沒有那麼單純

09 Kubernetes — Kustomize 是什麼？

9.1 Kustomize 在 Kubernetes 中的定位 ………………………… 9-2

9.2 Kustomize 介紹 ………………………………………………… 9-2

9.3　Kustomize 安裝 ···································· 9-4

9.4　基本指令 ··· 9-4

9.5　實戰演練 ··· 9-5

9.6　Kustomize 進階功能 ···························· 9-17

10 Kubernetes — 路由守護神 Ingress

10.1　Ingress 是什麼？ ······························ 10-2

10.2　Ingress 的作用 ······························· 10-3

10.3　Ingress 安裝 ································· 10-4

10.4　實戰演練 ····································· 10-6

11 Kubernetes — Pod 的生命週期

11.1　Pod 的生命週期 ······························ 11-2

11.2　Pod Phase（階段） ··························· 11-4

11.3　重啟策略（Restart Policy） ··················· 11-5

11.4　初始化容器（Init Container） ················· 11-6

11.5　生命週期鉤子（Lifecycle Hook） ··············· 11-8

11.6　健康檢查（Health Check） ···················· 11-9

12 Kubernetes Kubectl 指令與它的快樂夥伴

12.1　Kubectl 介紹 ································· 12-2

12.2　Kubectl 安裝設定 ···························· 12-3

12.3　Kubectl 語法 ································· 12-4

12.4　Kubectl 常用指令 ···························· 12-6

12.5　善加利用 Kubectl Help ······················ 12-11

Part 5　實戰部署篇：這些花式部署你學會了嗎？

13 Kubernetes Deployment Strategies — 常見的部署策略

13.1　重建部署（Recreate）………………………………… 13-2

13.2　滾動部署（Rolling Update）………………………… 13-4

13.3　藍綠部署（Blue / Green）…………………………… 13-5

13.4　金絲雀部署（Canary）………………………………… 13-7

13.5　A / B 測試（A / B Testing）………………………… 13-9

13.6　影子部署（Shadow）…………………………………… 13-10

14 Kubernetes Deployment Strategies — 重建部署與滾動部署

14.1　重建部署（Recreate）………………………………… 14-2

14.2　滾動部署（Rolling Update）………………………… 14-6

15 Kubernetes Deployment Strategies — 金絲雀部署

15.1　Nginx Ingress 金絲雀部署功能介紹 ………………… 15-2

15.2　金絲雀部署（Canary Deployment）………………… 15-3

15.3　使用金絲雀部署更新服務 …………………………… 15-4

15.4　實戰演練 ………………………………………………… 15-5

Part 6 主題篇—Volume：相較之下 Docker Volume 好像遜色了點？

16 Kubernetes Volume — Volume 是什麼？

16.1　那 Kubernetes 的 Volume 是什麼？ ………………… 16-2

16.2　Volume 類型 …………………………………………… 16-3

16.3　不同 Volume 的生命週期 …………………………… 16-5

17 Kubernetes Volume — EmptyDir

17.1　EmptyDir Volume ……………………………………… 17-2

17.2　實戰演練 ………………………………………………… 17-2

18 Kubernetes Volume — ConfigMap

18.1 ConfigMap 的特性 · 18-2

18.2 建立 ConfigMap · 18-3

18.3 實戰演練 · 18-9

19 Kubernetes Volume — Secret

19.1 什麼是 Secret？ · 19-2

19.2 建立 Secret · 19-3

19.3 實際應用 Secret · 19-6

19.4 聊聊關於 Secret 看起來並不那麼安全這件事 · · · · · · · · · · · · 19-7

20 Kubernetes Volume — PV & PVC

20.1 Storage Class · 20-2

20.2 Persistent Volumes（PV）· 20-3

20.3 Persistent Volume Claims（PVC）· · · · · · · · · · · · · · · · · 20-4

20.4 實戰演練 · 20-7

Part 7 主題篇—Resources：資源監控一定是全新的世界

21 Kubernetes Resources — Resource

21.1 Resource 是什麼？ · 21-2

21.2 Request 和 Limit 關係 · 21-3

21.3 Pod 的服務品質（Quality of Service，QoS）· · · · · · · · · · 21-4

21.4 Resource 設定的排列組合 · 21-7

21.5 實戰心得分享 · 21-8

22 Kubernetes Resources — Namespace

22.1 Namespace 是什麼以及何時使用？ · · · · · · · · · · · · · · · · · · 22-2

22.2 實戰演練 · 22-3

22.3 一些 Namespace 的特性 · 22-6

23 Kubernetes Resources — Resource Management

23.1 什麼是 LimitRange？ ···································· 23-2

23.2 什麼是 ResourceQuota？ ······························ 23-3

23.3 實戰演練 — LimitRange ····························· 23-5

23.4 實戰演練 — ResourceQuota ························· 23-9

24 Kubernetes Resources — Metrics Server

24.1 Metrics Server 是什麼？ ···························· 24-2

24.2 Metrics Server 原理 ······························· 24-4

24.3 安裝 Metrics Server ······························· 24-5

24.4 顯示資源使用訊息 ·································· 24-7

Part 8 主題篇—AutoScaling：身為 Server 守護者的你 是不是也沒辦法睡個好覺？

25 Kubernetes AutoScaling — AutoScaling 是什麼？

25.1 Autoscaler 的種類 ································· 25-2

25.2 Cluster Autoscaler（CA） ························· 25-3

25.3 Horizontal Pod Autoscaler（HPA） ··············· 25-4

25.4 Vertical Pod Autoscaler（VPA） ·················· 25-6

25.5 Multidimensional Pod Autoscaler（MPA） ········· 25-8

25.6 Custom Pod Autoscaler ·························· 25-10

26 Kubernetes AutoScaling — Horizontal Pod AutoScaler

26.1 確認 Metrics Server 是否就緒 ···················· 26-2

26.2 HPA 設定檔範例 ································· 26-3

26.3 實戰演練 ······································· 26-7

27 Kubernetes AutoScaling — Vertical Pod AutoScaler

27.1 確認 Metrics Server 是否就緒 ⋯⋯⋯⋯⋯⋯⋯⋯⋯⋯ 27-2

27.2 VPA 元件以及運作流程 ⋯⋯⋯⋯⋯⋯⋯⋯⋯⋯⋯⋯ 27-3

27.3 安裝 Custom Resource — VPA ⋯⋯⋯⋯⋯⋯⋯⋯⋯ 27-4

27.4 實戰演練 ⋯⋯⋯⋯⋯⋯⋯⋯⋯⋯⋯⋯⋯⋯⋯⋯⋯ 27-8

27.5 移除 VPA 模組 ⋯⋯⋯⋯⋯⋯⋯⋯⋯⋯⋯⋯⋯⋯⋯ 27-12

28 Kubernetes AutoScaling — Custom Pod AutoScaler

28.1 安裝 KEDA ⋯⋯⋯⋯⋯⋯⋯⋯⋯⋯⋯⋯⋯⋯⋯⋯ 28-2

28.2 KEDA（Kubernetes Event-Driven Autoscaling）⋯⋯⋯⋯⋯ 28-3

28.3 KEDA CRD—ScaledObject 和 ScaledJob ⋯⋯⋯⋯⋯ 28-4

28.4 KEDA 觸發器 ⋯⋯⋯⋯⋯⋯⋯⋯⋯⋯⋯⋯⋯⋯⋯ 28-7

28.5 KEDA 中的防抖動機制 Debouncing ⋯⋯⋯⋯⋯⋯⋯ 28-12

28.6 超越 Kubernetes HPA 的彈性伸縮 ⋯⋯⋯⋯⋯⋯⋯ 28-14

28.7 激活階段與縮放階段 ⋯⋯⋯⋯⋯⋯⋯⋯⋯⋯⋯⋯ 28-15

Part 9 主題篇—Security：朕不給的，你不能搶

29 Kubernetes Security — 使用 Context 進行用戶管理

29.1 Kubernetes 的認證與授權 ⋯⋯⋯⋯⋯⋯⋯⋯⋯⋯ 29-2

29.2 Kubernetes Context 是什麼？ ⋯⋯⋯⋯⋯⋯⋯⋯⋯ 29-3

29.3 用戶管理情境 ⋯⋯⋯⋯⋯⋯⋯⋯⋯⋯⋯⋯⋯⋯⋯ 29-5

29.4 實戰演練 ⋯⋯⋯⋯⋯⋯⋯⋯⋯⋯⋯⋯⋯⋯⋯⋯⋯ 29-6

29.5 所以說那個 Context 中的 Cluster 跟 User 呢？ ⋯⋯⋯⋯ 29-9

30 Kubernetes Security — RBAC Authorization 授權管理

30.1 深入了解 Kubernetes API Server ⋯⋯⋯⋯⋯⋯⋯⋯ 30-2

30.2 實戰使用 RBAC（Role-Based Access Control）⋯⋯⋯⋯ 30-4

30.3 Role vs ClusterRole ⋯⋯⋯⋯⋯⋯⋯⋯⋯⋯⋯⋯⋯ 30-10

30.4 RoleBinding vs ClusterRoleBinding ⋯⋯⋯⋯⋯⋯⋯ 30-12

Part 1

概念篇

· · · · · · · · · · · · ·

萬丈高樓平地起，
開始爬吧！

接下來的章節中，筆者將以自身學習經驗為讀者由淺入深地探討 Kubernetes。

Kubernetes 是什麼？

從本章開始，我們就要對 Kubernetes 有個初步的了解，俗話說有好的開始就是成功的一半，相信大家在接觸一個新領域時內心多少會有點痛苦跟懼怕，但了解後就會發現 Kubernetes 並沒有當初幻想的那麼艱深、那麼遙不可及，反而就像免治馬桶一樣用過就回不去。

Kubernetes 是一個相對進階的容器編排平台，需要具備基礎的容器化基礎，這裡可以簡單理解成：只要接觸過 Docker 一定程度的人應該都可以很快理解其中奧妙，所以我們將從基礎概念開始闡述 Kubernetes，並且開始由粗至細地去了解它的各個元件。

▶ 1.1 網路部署的演變

1. 傳統部署時代

早期的應用程式主要在物理伺服器上建構。由於物理伺服器無法限制運行應用程式的資源使用，這導致資源分配的問題，最初的解決方案是在不同的物理伺服器上運行每一個應用程式。除了物理伺服器的高昂維護成本，這種策略還存在一個問題：當某個伺服器的資源使用率不高時，其剩餘資源無法被其他應用程式利用。因此，人們開始思考是否可以在同一台物理伺服器上使用虛擬隔離技術，進而促成了下一個時代的技術演進。

2. 虛擬化部署時代

虛擬化技術逐漸被引入使用，允許在單一物理伺服器上運行多個虛擬環境（VM），這讓各種應用程式能在不同的 VM 之間安全地隔離。虛擬化技術讓物理伺服器能更有效地利用資源，並使應用程式更新更方便。此外，這種技術也降低了硬體成本。但從圖 1-1 可以看出，每個 VM 都有一個完整的操作系統，每次啟動都需要建立一個完整的系統環境，這同樣可能造成資源浪費。面對這個問題，現代的開發者提出了一種新的解決方案，即容器化技術，例如 Docker 和 Kubernetes。

3. 容器部署時代

容器與 VM 相似，但有更高的隔離彈性。各個容器能共享操作系統，只需安裝所需的應用程式即可，避免了不必要的資源浪費，實現了更輕量化的部署。同時，每個容器都可以擁有自己的檔案系統、CPU、記憶體、行程空間等。容器化是現代開發的主流趨勢，它帶來了以下好處：

- 快速的應用程式啟動和敏捷的部署：透過鏡像的不可變性，提供簡單的 Rollback 以及容器建構部署。
- 跨開發、測試和生產以及跨系統的一致性：在不同的設備上也可以擁有一致環境的應用程式。
- 提高抽象級別，實現低耦合、分散式的微服務：將應用程式分解成更小的單位，並且動態部署和管理，而不是在單體式架構下運行。
- 資源隔離：可預測的效能表現。
- 資源利用：高效率高密度。

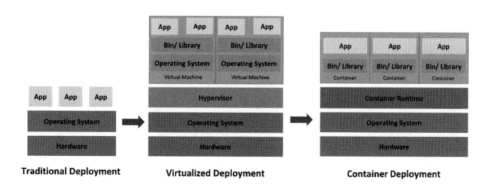

▲ 圖 1-1 網路部署的演變

▶ 1.2 Kubernetes 的定位

在容器化技術普及的時代，許多人已經將此技術作為他們的部署策略。然而，我們還需要管理和確保應用程式容器的狀態，以確保服務不會中斷。在這種情況下，Kubernetes 能提供一個可彈性運行分散式系統的框架，利用容器技術的敏捷部署和快速建構優點。在 Kubernetes 中，我們可以使用統一的設定檔來滿足擴展需求、實現故障轉移、自動部署、滾動更新和回復等操作，並進行監控和偵測。透過程式處理平台層的所有操作，我們甚至可以進一步實現 Infrastructure as Code（基礎設施即程式碼）。

筆者碎碎念

想到當初將應用程式打包成容器，並且在雲端服務上使用 Docker 成功運作起來的成就感，筆者不禁感嘆：天啊！雲端平台搭配 Docker 太方便了！但這種心得在我知道這世上存在著 Kubernetes 這種更宏觀的容器化技術，以及它的存在是要解決更實際的生產環境問題後，讓我知道自己還有很多不足的地方。

參考資料

- Kuebrnetes Overview
  https://kubernetes.io/zh-cn/docs/concepts/overview/

Kubernetes 的元件

從前一章的 Kubernetes 介紹，可以知道它是用於大規模運行分散式應用和服務的開源容器編排平台，在背後必須有一個穩固的結構來支持其運行，本章就來更深入認識組成 Kubernetes 的幾個大元件。

2.1 Kubernetes 設計原理

正如 Kubernetes 官方文件中所述，Kubernetes 叢集的設計基於 3 個原則。

Kubernetes 叢集應做到：

- 安全：它應遵循最新的安全最佳實踐。
- 易於使用：它應能透過一些簡單的指令進行操作。
- 可擴展：不應偏向於某一個提供商，而是能透過設定檔案進行自定義。

2.2 Kubernetes 叢集中包含哪些元件？

當你部署完 Kubernetes 時，便「至少」擁有了一個完整的叢集（可以依照需求擴展叢集），而每個叢集裡都會有個主節點（Master Node）作為控制整個叢集容器溝通，還有分派任務的控制平面元件（Control Plane Components），以及至少一個工作節點（Worker Node）託管所謂的 Pod，也就是作為應用負載的元件。

1. Node 節點

Node 是 Pod 真正運行的主機，是「Kubernetes 中最小的主機單位」，可以是物理機也可以是虛擬機，而叢集上的所有 Pod 將會被後面提到的 kube scheduler 分派到最合適的 Nodes 中（見 2-5 頁）。為了管理 Pod，每個 Node 節點上至少要運行 Container Runtime（比如 Docker）、kubelet 和 kube-proxy 服務。

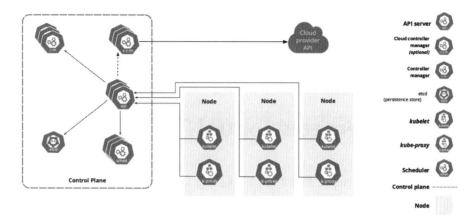

▲ 圖 2-1 Kubernetes 元件

2. Pod 容器集

Pod 是 Kubernetes 物件模型中最小、最基礎的單元。它代表了應用的單個實例。每個 Pod 都是由一個或一個以上的容器（每個容器對應一個 Image）以及若干控制器的組成。 Pod 跟 Docker 一樣支援持久儲存（Volume），以運行有狀態應用。

3. Container Runtime Engine 容器運行引擎

為了運行容器，每個 Node 都會有個容器運行引擎，像是 Docker ，但 Kubernetes 也支持其他符合開源容器運動（OCI）標準的運行引擎，例如 rkt 和 CRI-O。

4. Kubelet

每個計算節點中都包含一個 kubelet，這是一個與控制平面通訊的微型應用。kublet 可確保容器在容器集內運行。當控制平面需要在節點中執行某個操作時，kubelet 就會執行該操作。

5. Kube-proxy

每個 Node 中還包含 kube-proxy，這是一個用於優化 Kubernetes 網路服務的網路代理。kube-proxy 負責處理叢集內部或外部的網路通訊，靠操作系統的封包過濾層，或者自行轉發流量。可以簡單理解為負責為 Service 提供叢集內部的服務發現和負載均衡。

▷ 2.3 Kubernetes Control Plane

首先，來討論 Kubernetes 叢集的中樞神經—— Control Plane（控制平面）。在這裡，我們可以找到負責管理叢集的 Kubernetes 元件以及一些與叢集狀態和設定相關的資訊。這些核心元件扮演重要的角色，確保容器以所需的數量和資源正確運行。控制平面會持續與本地系統保持連線，並以核心方式運行，因此通常我們無需特別設定。

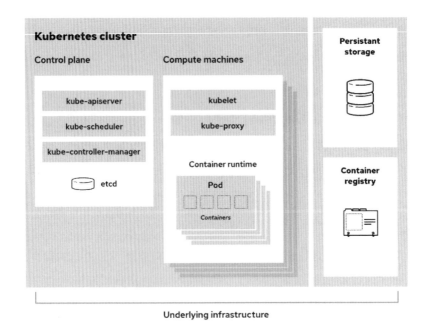

▲ 圖 2-2 Kubernetes Control Plane

1. kube-apiserver

kube-apiserver 是 Kubernetes 控制平面的主要元件之一,它提供了 Kubernetes API 的端點,允許使用者、叢集內部元件和外部服務進行互動。它處理 API 和驗證請求,然後執行相應的操作。這使得你能夠查詢叢集的狀態、改變叢集的狀態(例如建立、更新和刪除資源),以及訂閱特定資源的變更。

2. kube-scheduler

kube-scheduler 是負責將 Pod(Kubernetes 的最小部署單位)分配給合適的 Node 的元件。當你建立一個新的 Pod 或一組 Pod(例如透過 ReplicaSet、Deployment 等),kube-scheduler 會根據各種因素(包括叢集當前的資源使用狀態、Pod 的資源需求、調度策略和使用者提供的一些約束)來選擇最適合的 Node,例如 node affinity、taints 和 tolerations 等。

3. kube-controller-manager

kube-controller-manager 是 Kubernetes 控制平面的關鍵元件之一,它負責運行和管理集群中的各種控制器。它並不是將多個控制器整合為一個功能,而是在單一的行程中運行多個不同的控制器實例(例如節點控制器、端點控制器、ReplicaSet 控制器等),每個控制器都負責管理叢集中的特定方面。控制器持續監控叢集狀態,並在系統狀態與期望狀態不一致時,自動採取行動以修正這些差異。舉例來說,如果一個控制器監控 Pod 的數量,並且一個 Pod 失敗或結束,它將會試圖重新建立該 Pod 以符合預期的副本數量。kube-controller-manager 讓這些控制器能夠被集中管理和設定。

4. etcd

etcd 是一種分散式鍵值儲存系統,它用於保存 Kubernetes 的所有叢集資料。它是一個分散式的資料庫,使用 Raft 一致性算法確保資料一致性,從而提供高可用性和資料冗餘。在叢集的所有狀態(例如 Pod、Service、Volume 等的當前狀態和後設資料)都被保存在 etcd 中,並在需要時用於恢復叢集狀態。

筆者碎碎念

在簡單理解 Kubernetes 的結構後，可能對於其實際運用的概念還是一片茫然，但隨著之後的實戰演練後，會有很多機會回頭反覆驗證這個結構以及背後通訊的機制，然後漸漸內化成自己的理解，接下來我們將要實際操作 Kubernetes，開始親身體會它的靈活性以及自由度。

參考資料

- Kubernetes Cluster Architecture
  https://godleon.github.io/blog/Kubernetes/k8s-CoreConcept-Cluster-Architecture/

- Kubernetes Components
  https://kubernetes.io/docs/concepts/overview/components/

Part 2

安裝篇

一定要安裝些
什麼的吧！

工欲善其事，必先利其器，有了好用的工具，讓你事半功倍。

CHAPTER

03

安裝 Kubernetes
（Docker Desktop）

社群中有好幾種非常方便、可以讓開發者在本機輕易架設一個 Kubernetes 叢集的工具，本書將會以 Docker Desktop 內建 Kubernetes 進行安裝教學以及後面的本地操作。

眾所皆知 Kubernetes 的更新速度以及 API 棄用速度都是相當地快，大約每隔三個月就進行一次小更新，時常在網路上看到一年前左右的教學可能就已經沒辦法開箱即用了，這會讓我們必須不停地翻閱最新的 API 文件，甚至直接爬原始碼來轉換。因此，筆者希望這章節傳達的內容不只是手把手的教學，而是一通百通的觀念。

以下內容版本皆在 Docker Desktop 4.8.1 (78998) v20.10.14 中 Kubernetes v1.24.0 的 MAC 環境下建立。

3.1 下載 Docker Desktop（macOS）

STEP 1 前往官方下載連結：

▲ 圖 3-1 Docker 官方下載連結

（ https://dockerdocs.cn/docker-for-mac/install/#google_vignette ）

STEP 2 將 docker.dmg 拖曳至右邊的 Application。

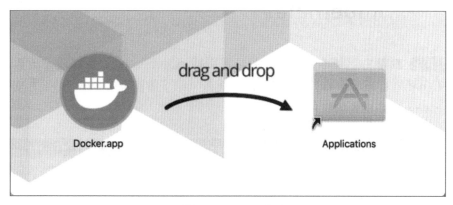

▲ 圖 3-2 Docker 拖曳

STEP 3 點擊 docker.app，看到圖 3-3 的初始介面就代表安裝成功了！

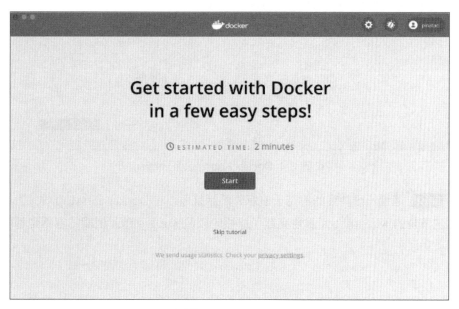

▲ 圖 3-3 Docker Start

▷ 3.2 開啟 Docker Desktop 內建的 Kubernetes

STEP 1 點擊右上角【setting】選項可以找到 Kubernetes 相關設定，再點選【Enable Kubernetes】並按下 Apply & Restart，這裡會花一小段時間在安裝 Kubernetes 所需的相關 Image。

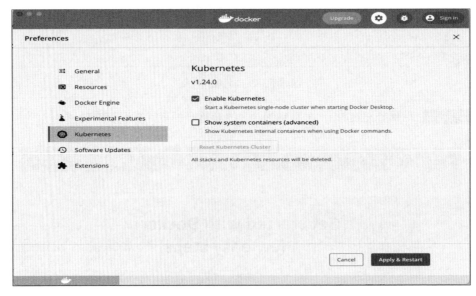

▲ 圖 3-4 Docker Kubernetes version

STEP 2 經過一段時間的等待，沒意外的話就可以在 Docker Desktop GUI 上看到左下角的 Kubernetes 服務亮起了綠燈，以及 Kubernetes 相關的容器也都順利啟動了。

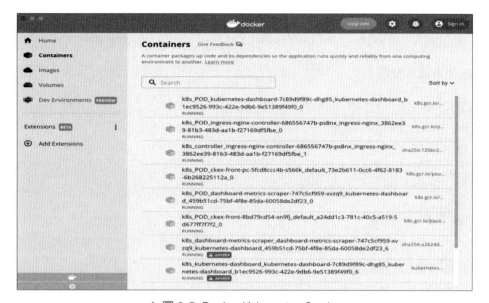

▲ 圖 3-5　Docker Kubernetes Service

⫸ 3.3 查看 Kubernetes 狀態

1. 查看 Cluster 資訊：

```
1.   kubectl cluster-info
2.   ---
3.   Kubernetes control plane is running at https://kubernetes.docker.
     internal : 6443
4.   CoreDNS is running at https://kubernetes.docker.internal : 6443/api/
     v1/namespaces/kube-system/services/kube-dns : dns/proxy
```

2. 查看 Nodes 資訊：

```
1.   kubectl get nodes
2.   ---
```

```
3.  NAME                STATUS      ROLES           AGE    VERSION
4.  docker-desktop      Ready       control-plane   60d    v1.24.0
```

3. 查看版本資訊：

```
1.  kubectl version --short
2.  ---
3.  Client Version: v1.24.0
4.  Kustomize Version: v4.5.4
5.  Server Version: v1.25.0
```

大功告成！打完收工。

 TIPS

從這裡我們就可以看出 kubectl 與我們使用指令與 Kuberentes 互動息息
相關，一般下載完 Docker Desktop 就會自動安裝。如果沒有安裝的話，
也可以直接使用 brew install kubectl 進行安裝。

筆者碎碎念

經過前幾章的初步了解，我們將要開始踏入 Kubernetes 的世界，在這邊也不禁感
慨，現在的後端已經很難將精力完全放在寫程式上。

為了處理好一件事情會使用到不同的後端語言，每個語言又有各自的 Web
Framework 或 API Framework，同時還可能會用上各種類型的資料庫，無論是關
聯式非關聯式又或者是快取型的資料庫。此外為了可以承受更大規模的使用量，
我們又必須要使用負載均衡、自動擴展以及資料庫主從分離等架構上的技術觀
念，管理這些服務伺服器時，我們不能單單地使用 Docker 容器化解決這些需求，
我們需要擁有一層更上層的平台容器管理層來幫我們把這件事情變得輕鬆，而這
正是我們需要學習 Kubernetes 的原因。

參考資料

- 在 Mac 上安裝 Docker Desktop
  https://dockerdocs.cn/docker-for-mac/install/#google_vignette

- Local Kubernetes for Mac– MiniKube vs Docker Desktop
  https://codefresh.io/blog/local-kubernetes-mac-minikube-vs-docker-desktop/

- Docker Desktop for Mac/Windows 開啟 Kubernetes
  https://github.com/AliyunContainerService/k8s-for-docker-desktop

安裝 **Kubernetes
Dashboard GUI**

在各種與容器以及服務設定的 Kubernetes 世界裡，我們會頻繁地使用 kubectl
指令去對 kube-apiserver 做請求並且回傳資訊到我們的終端機介面中，於是官
方就推出了一套 Kubernetes Dashboard 做為 Web UI 工具，不只可以一目了然
地列出所有容器的服務狀態，並且可以在上面將我們的 kubectl 指令轉變成在
UI 上的功能，可以說是在本地端使用 Kubernetes 的標配了（通常在雲端平台
會搭配雲端自家的監控整合介面）。

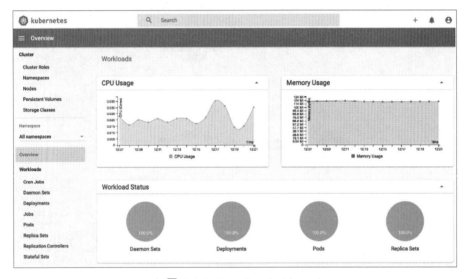

▲ 圖 4-1 kubernetes dashboard

4.1 Kubernetes Dashboard 是什麼？

Kubernetes Dashboard 是一種網頁式的使用者介面，讓使用者可以視覺化地查
看、控制和管理在 Kubernetes 叢集中運行的應用程式。透過 Dashboard，你不
只可以查看叢集及其資源的狀態，還可以對這些資源進行操作，例如建立、更
新和刪除 Kubernetes 資源（例如 Deployment、Job、DaemonSet 等），或視
覺化地監控 Pod 和 Node 的運行狀態。

Kubernetes Dashboard 除了提供基本的 Kubernetes 資源管理功能，還提供了許多強大的擴展功能，包括：

- 查看應用程式的狀態：Dashboard 提供了一個圖形化的介面，讓你可以快速查看應用程式的運行狀態，包括 Pod 的運行狀態、Deployment 的規模以及 Service 的資訊等。
- 監控和除錯：透過 Dashboard，你可以查看 Pod 的日誌，並在 Pod 出現問題時進行除錯。此外，Dashboard 還提供了與 Metrics Server 整合的應用程式監控功能，讓你可以視覺化地查看應用程式的 CPU 和記憶體使用狀況。
- 管理叢集資源：在 Dashboard 中，你可以對 Kubernetes 的各種資源進行管理，包括建立、更新和刪除資源。並且，你可以以將 YAML 或 JSON 格式的資源描述檔案直接上傳到 Dashboard 以方便建立或更新資源。

▶ 4.2 設定 Kubernetes Dashboard

1. 使用 Homebrew 安裝項目管理工具 Helm（macOS）：

```
1.   brew install helm
```

Helm 是一個管理 Kubernetes 應用的強大工具，礙於本書篇幅，筆者便不在此贅述。

2. 使用 Helm 下載 Kubernetes Dashboard 相關資源：

```
1.   helm repo add kubernetes-dashboard https://kubernetes.github.io/
     dashboard/
2.   helm upgrade --install kubernetes-dashboard kubernetes-dashboard/
     kubernetes-dashboard -n kubernetes-dashboard --set metricsScraper.
     enabled=true --create-namespace
```

3. 檢查 kubernetes-dashboard 應用狀態：

```
1.  kubectl get pod -n kubernetes-dashboard
2.  -------
3.  NAME                                        READY STATUS  RESTARTS AGE
4.  dashboard-metrics-scraper-748b4f5b9d-7vxjd  1/1   Running  0       26s
5.  kubernetes-dashboard-86b687bd84-wvf5q       1/1   Running  0       26s
```

4. 開啟 API Server 訪問代理並訪問 Kubernetes dashboard 頁面：

```
1.  export POD_NAME=$(kubectl get pods -n kubernetes-dashboard
    -l "app.kubernetes.io/name=kubernetes-dashboard,app.kubernetes.io/
    instance=kubernetes-dashboard" -o jsonpath="{.items[0].metadata.
    name}")
2.  echo http://localhost:8443/
3.  kubectl -n kubernetes-dashboard port-forward $POD_NAME 8443:8443
```

現在我們就可以透過 https://localhost:8443/ 訪問 Kubernetes Dashboard。

這時畫面將會出現需要訪問權限 token 的驗證：

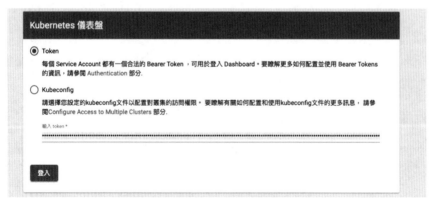

▲ 圖 4-2 kubernetes-dashboard login

5. 接下來我們需要產生一個預設 Service Account Token 來登入頁面。直接在終端機輸入下列指令：

```
1.  kubectl apply -f https://raw.githubusercontent.com/MikeHsu0618/
    kubernetes-from-another-world/main/ch4/default-sa.yaml
```

這時 Kubernetes 將會為我們產生擁有所需權限的 Service Account 以及 Token。

6. 印出 Token（macOS）：

```
1.  TOKEN=$(kubectl -n kube-system describe secret default | awk
    '$1=="token:"{print $2}')
2.  kubectl config set-credentials docker-desktop --token="${TOKEN}"
3.  echo $TOKEN
4.  // your token
5.  ...
```

7. 將上一步驟拿到的 Token 填入後，即可成功進入 Kubernetes Dashboard 主頁面！

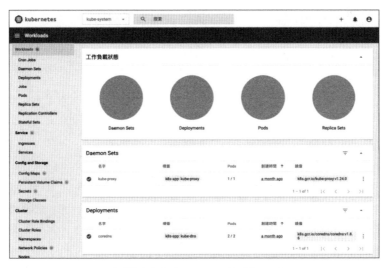

▲ 圖 4-3 kubernetes-dashboard

筆者碎碎念

Kubernetes Dashboard 不得不說是一個初學者以及指令苦手的救星，在初期對 Kubernetes 完全不熟悉的我，甚至連常用的指令都不清楚。有了 Kubernetes Dashboard 這個 GUI 幫助，讓我從另一個角度去更了解到 Kubernetes 的運作方式，同時也讓我學到更多相關指令，當然有很多唾棄 GUI 的指令神人會覺得圖形化工具會寵壞工程師，但莫忘世上苦人多，更多的是對較高的學習門檻打退堂鼓的平庸小白，因此我反而認為圖形化工具提供了一個機會，讓我們從另一個層面去學習一個工具，並且搭配指令操作就能相輔相成。

在簡單理解 Kubernetes 的結構後，讀者可能對其實際運用的概念還是一片茫然，但隨著之後的實戰演練，會有很多機會回頭反覆驗證這個結構以及背後通訊的機制，然後漸漸內化成自己的理解。接下來我們將要實際操作 Kubernetes，開始親身體會它的靈活性以及自由度。

參考資料

- Deploy and Access the Kubernetes Dashboard
  https://kubernetes.io/docs/tasks/access-application-cluster/web-ui-dashboard/

Part 3
基礎篇
....
老闆總說先可
以 **Run** 就好…

有人說學習游泳最快的方法就是跳進大海，就像實際工作時，老闆總對我們開發人員常說的一句話：「先可以 Run 就好，其他之後再說」。

Kubernetes — 實戰
做一個 Pod

當我們開始進行 Kubernetes 實作練習，並實現進階的操作，像是負載均衡、滾動更新、安全與監控等概念，我們都會不斷圍繞在 Pod、Service、Deployment 等資源單位的身邊，不停地為其設定來實現這些操作，因此，我們要先對這些資源進行初步的認識。

首先要介紹的是 Pod ，在第二章的 Kubernetes 元件介紹中也有提到，Pod 是 Node（節點）中最小的工作負載單位，等於我們所使用的容器們都會被放置到 Pod 中管理。通常一個 Pod 裡只會有一個容器（也可以有一個以上的容器），接下來我們將會實作一個簡單的練習範例。

▶ 5.1 建立容器（Container）

基本上只要是 Container-Based 的服務，都可以拿來讓 Kubernetes Pod 部署，而且不僅限於 Docker Container，只是因為現代容器趨勢依舊是被 Docker 掌握著，加上它的方便性以及穩定性，所以我們將會使用 Docker 處理各種容器建構。

NOTE

後面將會使用 Golang 程式碼建立出一個簡易的 API Server 的 Image，以模擬實際工作中部署 Kubernetes 服務時會面對的各種場景。請讀者使用熟悉語言建立自己的 Image，幫助自己運用在符合自身的應用情境，或直接翻閱到下個段落，使用本章節提供的範例 Image 即可。

---

1. Golang API Server 範例程式碼

以下是一個簡單 Golang API Server 範例：

範例檔 **main.go**

```
1.  # main.go
2.  package main
3.
4.  import (
5.      "github.com/gin-gonic/gin"
6.      "net/http"
7.  )
8.
9.  func main() {
10.     router : = gin.Default()
11.     router.GET("/", func(c *gin.Context) {
12.         c.JSON(http.StatusOK, gin.H{"data": "Hello foo"})
13.     })
14.     router.Run()
15. }
```

熟悉 Golang 的朋友，大概可以簡單的看出輸入 go run main.go 運行起這個 main.go 服務後，就是一個在預設 8080 port 號下的 localhost:8080/ 會回傳 Hello foo 字樣的簡易 API Server。

2. Dockerfile 設定檔

現在我們就來將這個 API Server 包成 Docker Image：

範例檔 **dockerfile**

```
1.  # dockerfile
2.  FROM golang: 1.18-rc-alpine as builder
3.  WORKDIR /
```

```
4.  COPY . .
5.  RUN go mod tidy
6.  RUN go build -o main
7.
8.  # docker mult sage build
9.  FROM alpine : 3.15.0-rc.4
10. WORKDIR /
11. COPY --from=builder /main .
12. EXPOSE 8080
13. ENTRYPOINT ["./main"]
```

以上就是我們簡單產生出一個 Docker Image 的設定檔，接下來我們就可以用它來建構 Image 與 Container。

```
1.  docker build -t foo .
```

馬上來執行 docker run 啟動 API Server 看看：

```
1.  docker run --rm -p 8080:8080 -it -d foo
```

輸入 localhost:8080，會看到成功回傳 Hello foo 字樣：

▲ 圖 5-1 api response

3. 將 Image 推送到 Container Registry

當我們將映像檔推送到遠端的映像檔儲存服務的倉庫時，我們便能以映像檔和標籤方式來取得我們所需的特定映像檔，這是採用容器化的便利性之一。當然，Kubernetes 也是以同樣的方式來取得要運行的容器服務。

推上 Docker Hub：

```
1.   # build and tag
2.   docker build -t mikehsu0618/foo .
3.   # push to registry
4.   docker push mikehsu0618/foo
```

這裡簡單地將 Docker Image 推到我個人的 Docker Hub 庫中，Docker 相關操作就不在此贅述，如需練習，可以直接拉取這裡已經建好的 Image。

⑉ 5.2 建立 Kubernetes 設定檔

接下來我們將要試著把這個服務用 Kubernetes 部署起來！首先我們需要本地建立設定檔，讓我們接下來的指令去運行到指定檔案。

範例檔 **pod.yaml**

```
1.   # pod.yaml
2.   apiVersion: v1
3.   kind: Pod
4.   metadata:
5.    name: foo
6.    labels:
7.      app: foo
8.   spec:
9.    containers:
```

```
10.      - name: foo
11.        image: mikehsu0618/foo
12.        ports:
13.          - containerPort: 8080
```

- apiVersion：代表目前該元件在 Kubernetes 中的使用的 API 版本。
- metadata.name：表示這個 Pod 資源的名稱。
- metadata.labels：Label 是附加在 Kubernetes 物件上的標籤，用於對使用者有意義且相關的物件進行分組，但它們並不直接影響核心系統的運作。有了 Label，我們就可以利用選擇器（Selector）來挑選具有特定 Label 的物件。
- spec：在 spec 我們可以看到在 Docker 中很熟悉類似 Image 的設定，此處可以用來設定一個或多個 Container。
- containers.name：設定 Container 名稱。
- containers.image：Image 的路徑，此處為 Docker Hub 的鏡像拉取路徑。
- containers.ports：containerPort 代表的是這個 Container 開放的 port 來允許外部資源存取。所以這邊根據我們的應用程式以及 dockerfile 設定的 8080 port 來指定 8080 連接埠。

⑉ 5.3 在 Kubernetes 中建立 Pod

1. 通常在 Kubernetes 中，所有指令的建立都可以用這兩種指令來完成：

```
1.  # 使用 -f 參數去指定設定檔路徑並且建立 pod
2.  kubectl apply -f pod.yaml
3.  # or
4.  kubectl create -f pod.yaml
```

以上兩個指令都可以用來建立資源，差別在於 create 只能用來建立還未存在的資源，而 apply 可以資源在已經存在的情況下，查看設定是否有異動並且更新，兩者使用情境上有所區別。

2. 接著查看 Pods 列表：

```
1.  kubectl get pods
2.  -------
3.  NAME            READY   STATUS    RESTARTS    AGE
4.  foo             1/1     Running   0.          26s
```

並且可以使用 kubectl describe 指令查看 Pod 更詳細的資訊：

```
1.  kubectl describe pod foo
2.
3.  ==========================
4.  foo:
5.      Container ID: docker://fc02b6801a4b6c62cb0aed77f4480bb41f297476
    c034a0d75fa079ea7e883cb4
6.      Image:          mikehsu0618/foo
7.      Image ID:       docker-pullable://mikehsu0618/foo@sha256 :
    41cd860bd4c9ce86271bb5c864599bb2ca32b3c3a377afb12d2142fa12a87b1d
8.      Port:           8080/TCP
9.      Host Port:      0/TCP
10.     State:          Running
11.       Started:      Tue, 21 Jun 2022 17: 13: 30 +0800
12.     Ready:          True
13.     Restart Count:  0
14.     Environment:    <none>
15.     Mounts:
16.       /var/run/secrets/kubernetes.io/serviceaccount from kube-api-
    access-v84zl (ro)
17. Conditions:
18.   Type              Status
19.   Initialized       True
20.   Ready             True
21.   ContainersReady   True
```

```
22.    PodScheduled        True
23. Volumes:
24.    kube-api-access-v84zl:
25.     Type:        Projected (a volume that contains injected data from
    multiple sources)
26.     TokenExpirationSeconds:  3607
27.     ConfigMapName:        kube-root-ca.crt
28.     ConfigMapOptional:      <nil>
29.     DownwardAPI:          true
30. QoS Class:               BestEffort
31. Node-Selectors:          <none>
32. Tolerations:             node.kubernetes.io/not-ready :
                             NoExecute op=Exists for 300s
33.    node.kubernetes.io/unreachable : NoExecute op=Exists for 300s
34. Events:
35.   Type    Reason       Age    From                   Message
36.   ----    ------       ----   ----                   -------
37.   Normal  Scheduled    2m37s  default-scheduler  Successfully
    assigned default/foo to docker-desktop
38.   Normal  Pulling      2m37s  kubelet Pulling image "mikehsu0618/foo"
39.   Normal  Pulled       2m33s  kubelet Successfully pulled image
    "mikehsu0618/foo" in 3.777043918s
40.   Normal  Created      2m33s  kubelet  Created container foo
41.   Normal  Started      2m32s  kubelet  Started container foo
```

5.4 使用 kubectl port-forward 與 Local 端接軌

在 Kubernetes 的世界中，許多資源和服務都是運行在內部的，並不能直接從我們的本地機器進行訪問。這種情況下，如何才能直接和這些服務進行互動，並獲得我們所需的資訊或進行測試呢？

這就是 kubectl port-forward 能派上用場的地方。利用 kubectl port-forward，我們能夠將 Kubernetes 中的服務連接埠映射到本地連接埠，從而可以直接從我們的本地環境訪問到該服務。無論是要查看一個在 Kubernetes 叢集內部運行的應用日誌，還是需要在本地進行除錯，kubectl port-forward 都能提供一個非常方便的方法來實現這一目標。

因此，對於任何使用 Kubernetes 並希望能直接與叢集內部的服務進行互動的開發者來說，了解和掌握 kubectl port-forward 的使用都是非常重要的。在接下來的內容中，我們將會具體介紹如何使用 kubectl port-forward，並實現 Local 端和 Kubernetes 服務的連接！

來看一下 kubectl port-forward 指令：

```
1.   kubectl port-forward TYPE/NAME [options] [LOCAL_PORT: ]REMOTE_PORT
```

port-forward 可以使我們本地的連接埠轉發射到指定的 Kubernetes 叢集連接埠中，這麼一來我們就可以使用本地的 localhost 轉發到我們建立起來的 Kubernetes 服務。

於是我們讓啟動的 Pod 的 8080 連接埠轉發到本地的 8080 連接埠：

```
1.   kubectl port-forward pod/foo 8080:8080
2.   =====================================
3.   Forwarding from 127.0.0.1: 8080 -> 808080
4.   Forwarding from [: : 1] : 8080 -> 8080
```

沒意外的話我們使用 curl 去存取 localhost:8080 可以得到預期的回覆：

```
1.   curl http://localhost:8080
2.   -------
3.   {"data": "Hello foo"}
```

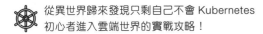
筆者碎碎念

如此一來我們已經學會使用 Kubernetes 運行了第一個容器，到後面隨著使用情境的複雜化，要處理的問題也會越來越難，但所有的操作不外乎完全圍繞在 Pods 身上，可以說是非常核心的觀念，在初期也是會不斷的與之接觸。

不難發現，我們用 Kubernetes 模擬了日常 Docker 流的一個 Container 使用方式，相信平常就有在用 Docker 的人會感到非常親切，可以說 Docker 是容器化技術的入門磚也不過分。

Kubernetes — 實戰 做一個 Service

還記得前一章我們使用 port-forward 實現將本地的接口轉發到 Kubernetes 中指定的 Pod 操作嗎？而 Service 元件就是 Kubernetes 特地為了用來定義「一群 Pod 要如何被連線及存取」的元件，此舉不只將連接埠的暴露任務進行解耦和抽象，利用 Labels 和 Selectors 更有效地識別哪些非永久資源的 Pod 需要應用該設定，如此一來，當 Pod 被動態的建立或銷毀時，相較於直接使用 port-forward 的 Pod 將會失去它的暴露連接埠，而新建立的 Pod 會保留原來的 Labels，確保我們之前的暴露連接埠設定繼續被套用。

因此，使用 Service 元件可以讓我們更有效地管理 Kubernetes 中的 Pod 連接和存取，並提供更靈活的設定選項，以應對 Pod 的動態變化。

▶ 6.1 Service 是什麼？

Service 是 Kubernetes 內定義的抽象化物件（Object），官方網站的介紹傳神地描述它的基本（原始）用途。

> A Kubernetes Service is an abstraction which defines a logical set of Pods and a policy by which to access them.
> Kubernetes Service 是個抽象化的概念，主要定義了邏輯上的一群 Pod 以及如何存取他們的規則。

▶ 6.2 那什麼是邏輯上的一群 Pod ？

每個 Pod 本身會帶著一或多個不等的標籤在身上，當 Service 指定存取某些特定的標籤時，Label Selector 會根據 Pod 身上自帶的標籤進行分組，並回傳 Service 所要求的 Pod 資訊。

▍用一句話簡單描述，就是帶著相同標籤、做類似事情的一群 Pod。

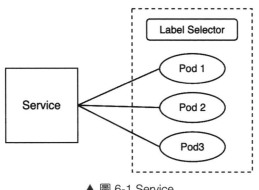

▲ 圖 6-1 Service

Service 作為中介層，避免使用者和 Pod 進行直接連線，除了讓我們的服務維持一定彈性，能夠選擇不同的 Pod 來處理請求之外，某種程度上亦避免裸露無謂的 Port 而導致資安問題。

另外，也體現出雲端服務架構設計中一個非常重要的觀念：

「**對於使用者而言，僅須知道有人會處理他們的請求，而毋須知道實際上處理的人是誰。**」

1. 使用 Service 實現 Load Balance

讓我們將前面的 foo 再新增一個夥伴 bar，並且使用 Service 來實作看看如何管理這兩個 Pod。

範例檔 **pod.yaml**

```
1.  # pod.yaml
2.  apiVersion: v1
3.  kind: Pod
4.  metadata:
5.    name: foo
```

```
6.    labels:
7.      app: foo
8.      type: demo
9.  spec:
10.   containers:
11.    - name: foo
12.        image: mikehsu0618/foo
13.        ports:
14.          - containerPort: 8080
15. ---
16. apiVersion: v1
17. kind: Pod
18. metadata:
19.   name: bar
20.   labels:
21.     app: bar
22.     type: demo
23. spec:
24.   containers:
25.    - name: bar
26.        image: mikehsu0618/bar
27.        ports:
28.          - containerPort: 8080
```

2. 撰寫 Service 設定

範例檔 service.yaml

```
1.  # service.yaml
2.  apiVersion: v1
3.  kind: Service
4.  metadata:
5.    name: my-service
```

```
6.  spec:
7.    selector:
8.      type: demo
9.    type: LoadBalancer
10.   ports:
11.    - protocol: TCP
12.      port: 8000
13.      targetPort: 8080
```

- apiVersion：此 Service 使用 Kubernetes API 的 v1 版本。

- metadata.name：此為該 Service 的名稱，用於 Kubernetes 中的唯一識別。

- spec.type：此處用來指定 Service 的型別，可選為 NodePort 或 LoadBalancer。NodePort 是以每個 Node 的 IP 來對外暴露服務，而 LoadBalancer 通常與雲端服務提供商的負載均衡器結合使用。

- spec.ports.port：此處用來指定建立的 Service 的 Cluster IP 將透過哪個 port number 來對應到 targetPort。

- spec.ports.nodePort：此處用以指定 Node 物件將透過哪一個 port number 來對應到 targetPort。如果在 Service 的設定檔中沒有明確指定，則 Kubernetes 會隨機選擇一個 port number。

- spec.ports.targetPort：這是我們指定的 Pod 要使用的 port number。由於我們的 Pod 中運行了一個在 port 8080 的 container (foo & bar)，我們將特定的 port number 指定到 hello-service，以便所有的流量都可以導向該 container。

- spec.ports.protocol：目前 Service 支援 TCP、SCTP 與 UDP 三種 protocol，預設為 TCP。每種 protocol 有其特定的使用場景和需求。

- spec.selector：selector 用於選擇標籤為 type=demo 的 Pod。在我們的範例中，建立的 Service 會將特定 port number 收到的流量導向這些 Pod。

3. 運行 Service 和 Pod 設定檔

```
1.   kubectl apply -f pod.yaml,service.yaml
```

查看服務狀態：

```
1.   # 查看所有服務
2.   kubectl get services
3.   # 查看全部元件狀態
4.   kubectl get all
```

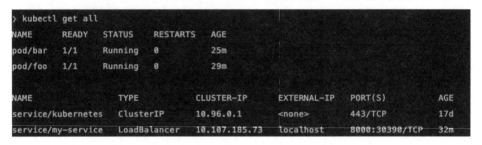

▲ 圖 6-2 all services

由圖 6-2 可以看出，我們順利 run 起一個 LoadBalancer。如果我們的 Kubernetes Cluster 是部署在第三方雲端服務（Cloud Provider），例如 Amazon 或 Google Cloud Platform，可以透過這些 Cloud Provider 提供的 LoadBalancer，幫我們分配流量到每個 Node，而我們使用 docker-desktop 直接會預設幫我們把 External IP 指向到我們的 http://localhost。

由此一來，當我們使用 curl 去 call 我設定好的 8000 連接埠，Kubernetes Service 就會將流量隨機分配到 foo bar 這兩個 container 中。

```
> curl http://localhost:8000/
{"data":"Hello bar"}%
> curl http://localhost:8000/
{"data":"Hello foo"}%
> curl http://localhost:8000/
{"data":"Hello bar"}%
> curl http://localhost:8000/
{"data":"Hello foo"}%
> curl http://localhost:8000/
{"data":"Hello bar"}%
> curl http://localhost:8000/
{"data":"Hello foo"}%
> curl http://localhost:8000/
{"data":"Hello foo"}%
> curl http://localhost:8000/
{"data":"Hello foo"}%
> curl http://localhost:8000/
{"data":"Hello bar"}%
> curl http://localhost:8000/
{"data":"Hello foo"}%
> curl http://localhost:8000/
{"data":"Hello bar"}%
> curl http://localhost:8000/
```

▲ 圖 6-3 curl localhost

🗓 筆者碎碎念

Kubernetes 的 Service 讓我們非常簡單地實現管理 Pod 流量以及 LoadBalance
的功能,這在以前可是需要到雲端平台一個一個設定才可以辦到。Kubernetes 幫
我們省下許多細節,使我們可以專注在實現維運部署的邏輯上面,但背後的觀念
非常值得我們回來細細咀嚼,在這裡只是先做一個簡單的小 demo,日後將會對
Service 的運作機制做更深入的了解。

參考資料

- Kubernetes Service 深度剖析 - 標籤對於 Service 的影響
  https://tachingchen.com/tw/blog/kubernetes-service-in-detail-2/

- Kubernetes Service 概念詳解
  https://tachingchen.com/tw/blog/kubernetes-service/

- Kubernetes Docs – Service
  https://kubernetes.io/zh-cn/docs/concepts/services-networking/
  service/#protocol-support

CHAPTER

07

Kubernetes — 實戰
做一個 Deployment

本章要介紹的是 Kubernetes 三兄弟的 Deployment，這個資源物件為 Pod
和 ReplicaSet 兩者提供了一個宣告式（Declarative）定義的方法來達到使用
者所期望的容器執行狀態，並且官方建議透過 Deployment 來部署 Pod 和
ReplicaSet，典型的應用場景包括：

● 定義 Deployment 來建立 Pod 和 ReplicaSet
● 滾動升級和回滾應用
● 擴展和收縮
● 暫停和繼續 Deployment

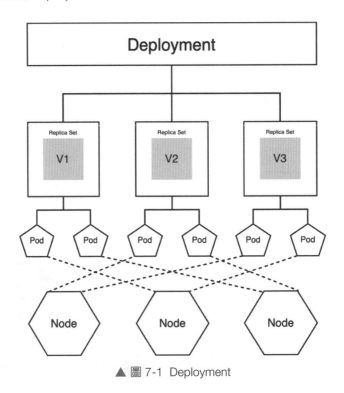

▲ 圖 7-1 Deployment

Pod 的介紹相信大家已經都不陌生了，但這邊怎麼又冒出一個 ReplicaSet
呢？ ReplicaSet 的主要目的是確保指定數量的 Pod 副本始終在運行。它確保
叢集中運行的 Pod 數量與使用者期望的數量（Desired Status）相匹配。儘管

ReplicaSet 可以獨立使用，但官方建議將其與 Deployment 一起使用，因為 Deployment 提供了更高級的功能和更好的管理體驗。

從圖 7-1 可以看出三者在 Kubernetes 中的對應關係。

▷ 7.1 使用案例

官方貼心地為我們提供了幾個經典的 Deployment 使用案例：

1. 使用 Deployment 來建立 ReplicaSet，而 ReplicaSet 在後台建立 Pod 並檢查成功或失敗。
2. 更新 Deployment 的 Pod 設定來宣告 Pod 的新狀態。這會建立一個新的 ReplicaSet，Deployment 將會按照控制速率（Controlled Rate）將 Pod 狀態更新至新的 ReplicaSet 設定。
3. 回滾到先前的 Deployment 版本，如果當前的版本不穩定。
4. 擴展或收縮 Deployment 以承載更多負荷。

接下來我們將用以上情境來實戰演練一下。

▷ 7.2 實戰演練

1. 建立 Deployment

範例檔 **deployment.yaml**

```
1.  # deployment.yaml
2.  apiVersion: apps/v1
3.  kind: Deployment
4.  metadata:
```

```
5.      name: foo-deployment
6.      labels:
7.        type: demo
8.  spec:
9.    replicas: 1
10.   selector:
11.     matchLabels:
12.       type: demo
13.   template:
14.     metadata:
15.       labels:
16.         type: demo
17.     spec:
18.       containers:
19.         - name: foo
20.           image: mikehsu0618/foo
21.           ports:
22.             - containerPort: 8080
23. ---
24. apiVersion: apps/v1
25. kind: Deployment
26. metadata:
27.   name: bar-deployment
28.   labels:
29.     type: demo
30. spec:
31.   replicas: 1
32.   selector:
33.     matchLabels:
34.       type: demo
35.   template:
36.     metadata:
```

```
37.        labels:
38.          type: demo
39.      spec:
40.        containers:
41.        - name: bar
42.            image: mikehsu0618/bar
43.            ports:
44.             - containerPort: 8080
```

重要參數：

- kind：kind 選擇我們使用的資源種類 Deployment。
- spec.replicas：被選擇套用的 Container 需要產生多少個 Pod，也是實現水平擴展的關鍵。
- spec.selector.matchLabels：這個欄位定義了哪些 Pod 應該被這個 Deployment 管理。它會選擇所有匹配這些標籤的 Pod。
- spec.template.metadata.labels：設定 template.spec 的 Label。
- spec.template.spec.containers：這個欄位包含了我們希望在 Pod 中運行的容器的設定。

接著讓我們運行設定：

```
1.   kubectl apply -f ./deployment.yaml
2.   --------------------
3.   deployment.apps/foo-deployment created
4.   deployment.apps/bar-deployment created
```

使用 kubectl get all 指令確認：

```
1.   kubectl get all
2.   --------------------
3.   NAME                                   READY   STATUS    RESTARTS   AGE
4.   pod/bar-deployment-75bcfbd655-g5gwm    1/1     Running   0          5m59s
```

```
5.   pod/foo-deployment-6bbf665b47-kfvxr  1/1   Running  0        5m59s
6.
7.   NAME                   TYPE        CLUSTER-IP  EXTERNAL-IP PORT(S)  AGE
8.   service/kubernetes    ClusterIP  10.96.0.1    <none>      443/TCP  23d
9.
10.  NAME                            READY  UP-TO-DATE  AVAILABLE  AGE
11.  deployment.apps/bar-deployment  1/1    1           1          5m59s
12.  deployment.apps/foo-deployment  1/1    1           1          5m59s
13.
14.  NAME                                   DESIRED  CURRENT  READY  AGE
15.  replicaset.apps/bar-deployment-75bcfbd655  1    1        1      5m59s
16.  replicaset.apps/foo-deployment-6bbf665b47  1    1        1      5m59s
```

看到我們成功的運行起了 foo bar 兩個 Pod，並且建立了各自的 Deployment
ReplicaSet。

7.3 更新 Deployment 實現水平擴展

接下來我們使用來使用不同的方法更新已經運行起來的 Deployment。

直接修改原有的設定檔：

```
1.   apiVersion: apps/v1
2.   kind: Deployment
3.   metadata:
4.     name: foo-deployment
5.     labels:
6.       type: demo
7.   spec:
8.     # 這裡我們將 Pod Replicas設定為 2！
9.     ====================
```

```
10.    replicas: 2
11.    ====================
12.    selector:
13.      matchLabels:
14.        type: demo
15.    template:
16.      metadata:
17.        labels:
18.          type: demo
19.      spec:
20.        containers:
21.          - name: foo
22.            image: mikehsu0618/foo
23.            ports:
24.              - containerPort: 8080
25. ---
26. # 不調整 bar-deployment
27. …
```

修改完後再執行一次 apply 指令，kubectl 會檢查指定設定檔是否有更新：

```
1.   # --record 可以記錄 rollout 歷史變更指令
2.   kubectl apply -f ./deployment.yaml --record
3.   ------------------------
4.   deployment.apps/foo-deployment configured   # 有更新
5.   deployment.apps/bar-deployment unchanged      # 未檢查到更新
```

接著可以使用 kubectl rolloout status 查看我們對 foo-deployment 的資源管理
狀態：

```
1.   kubectl rollout status deployment foo-deployment
2.   ------------------------
3.   deployment "foo-deployment" successfully rolled out
```

當指令顯示成功，即代表剛剛的更新已經正式生效，但只要遇到設定錯誤或者是無法實現的請求時，rollout status 將會持續等待至 timeout。

我們也可以使用第二個方法「指令更新」來調整 Deployment：

```
1.  kubectl scale deployment bar-deployment --replicas 3
```

而第三個方法為直接編輯在 Kubernetes 運行中的 Deployment 設定：

```
1.  // 打開 command 編輯面板，直接修改 replicas 數量
2.  kubectl edit deploy bar-deployment
```

來使用 kubectl get all 確認看看吧：

```
1.  kubectl get all
2.  ------------------------
3.  NAME                                      READY  STATUS   RESTARTS AGE
4.  pod/bar-deployment-75bcfbd655-75qcd       1/1    Running  0        31s
5.  pod/bar-deployment-75bcfbd655-c5h9w       1/1    Running  0        31s
6.  pod/bar-deployment-75bcfbd655-g5gwm       1/1    Running  0        5h52m
7.  pod/foo-deployment-6bbf665b47-45c2k       1/1    Running  0        4h7m
8.  pod/foo-deployment-6bbf665b47-kfvxr       1/1    Running  0        5h52m
9.
10. NAME                  TYPE       CLUSTER-IP   EXTERNAL-IP  PORT(S)   AGE
11. service/kubernetes    ClusterIP  10.96.0.1    <none>       443/TCP   23d
12.
13. NAME                             READY  UP-TO-DATE  AVAILABLE  AGE
14. deployment.apps/bar-deployment   3/3    3           3          5h52m
15. deployment.apps/foo-deployment   2/2    2           2          5h52m
16.
17. NAME                                        DESIRED  CURRENT  READY AGE
18. replicaset.apps/bar-deployment-75bcfbd655   3        3        3     5h52m
19. replicaset.apps/foo-deployment-6bbf665b47   2        2        2     5h52m
```

我們在回傳結果中可以看到 pod/bar-deployment 已經預期地啟動三個，並且 RepolicaSet 和 Deployment 也更新了對應狀態。

⯮ 7.4 使用 Rollout 查看歷史版本並回滾

在我們更新 Deployment 時，Kubernetes 會產生一個 Deployment Revision，可以簡單理解為是更新歷史版本，但要注意不是每一次的更新都會產生 Revision，只有在 Deployment created 以及 spec.template 範圍下的設定有更新才會產生，所以我們前面更新的 replicas=3 並不會出現在歷史中。

讓我們改動 spec.template 來實驗看看：

```
1.  apiVersion: apps/v1
2.  kind: Deployment
3.  metadata:
4.    name: bar-deployment
5.    labels:
6.      type: demo
7.  spec:
8.    replicas: 3
9.    selector:
10.     matchLabels:
11.       type: demo
12.   # 只有在 spec.template 下的改定才會記錄在 rollout history 中
13.   template:
14.     metadata:
15.       labels:
16.         type: demo
17.     spec:
18.       containers:
```

```
19.     - name: bar
20.         // 將我們的 image tag 版號改成不存在 'v1'
21.         ===================================
22.         image: mikehsu0618/bar: v1
23.         ===================================
24.       ports:
25.         - containerPort: 8080
```

更新 Deployment 設定檔並使用 --record 來記錄指令：

```
1.  kubectl apply -f deployment.yaml --record
2.  ------------------------
3.  Flag --record has been deprecated, --record will be removed in the
    future
4.  deployment.apps/foo-deployment unconfigured
5.  deployment.apps/bar-deployment configured
```

原本的 spec.template 雖然會被記錄在 rollout history 中，但不會有額外資訊，
--record 可以讓 Kubernetes 幫我們記下我們當下改變設定的那個指令。

這時我們就能在 rollout history 查看產生出來的 revision：

```
1.  kubectl rollout history deployment bar-deployment
2.  ------------------------
3.  REVISION   CHANGE-CAUSE
4.  1          <none>
5.  2          kubectl apply --filename=deployment.yaml --record=true
```

第一個版本為先前 Deployment 被建立時且沒有輸入 --record 的版本，第二個
版本為我們調整 bar image=mikehsu0618/bar:v1 且有 --record 的版本。

指定 revision 並查看詳細資訊：

```
1.  kubectl rollout history deployment bar-deployment --revision=2
2.  ------------------------
```

```
3.   deployment.apps/bar-deployment with revision #2ment --revision=2
4.   Pod Template:
5.     Labels:        pod-template-hash=864b65d8b6
6.          type=demo
7.     Annotations:  kubernetes.io/change-cause: kubectl apply
     --filename=deployment.yaml --record=true
8.     Containers:
9.      bar:
10.      Image:       mikehsu0618/bar: v1
11.      Port:        8080/TCP
12.      Host Port:   0/TCP
13.      Environment:        <none>
14.      Mounts:      <none>
15.    Volumes:       <none>
```

接下來一樣是使用 get all 指令查看容器狀況：

```
1.   kubectl get all
2.   -------------------------
3.   NAME                                    READY  STATUS          RESTARTS  AGE
4.   pod/bar-deployment-75bcfbd655-5b9z5     1/1    Running         0         7m5s
5.   pod/bar-deployment-75bcfbd655-6whzr     1/1    Running         0         7m5s
6.   pod/bar-deployment-75bcfbd655-zk88d     1/1    Running         0         7m5s
7.   pod/bar-deployment-864b65d8b6-wdhz9     0/1    ImagePullBackOff 0        6m34s
8.   pod/foo-deployment-6bbf665b47-dhndq     1/1    Running         0         40m
9.   pod/foo-deployment-6bbf665b47-pnjfs     1/1    Running         0         40m
10.
11.  NAME                  TYPE       CLUSTER-IP    EXTERNAL-IP   PORT(S)    AGE
12.  service/kubernetes    ClusterIP  10.96.0.1     <none>        443/TCP    23d
13.
14.  NAME                                READY  UP-TO-DATE  AVAILABLE  AGE
15.  deployment.apps/bar-deployment      3/3    1           3          7m6s
16.  deployment.apps/foo-deployment      2/2    2           2          40m
```

```
17.
18.  NAME                                             DESIRED   CURRENT   READY  AGE
19.  replicaset.apps/bar-deployment-75bcfbd655        3         3         3      7m6s
20.  replicaset.apps/bar-deployment-864b65d8b6        1         1         0      6m34s
21.  replicaset.apps/foo-deployment-6bbf665b47        2         2         2      40m
```

這時我們會發現我們的 pod/bar-deployment 發生了 ImagePullBackOff ，原因是我們並沒有建立 mikehsu0618/bar: v1 的 image ，這種情況很好地提供我們一個因為「推進到一個不穩定的版本」而需要使用版本回滾，復原服務到上一個正常的版本。

使用 rollout 的回滾指令復原先前版本設定：

```
1.   # 回滾至上個版本
2.   kubectl rollout undo deployment bar-deployment
3.
4.   # 回滾至指定版本
5.   kubectl rollout undo deployment bar-deployment --to-revision=1
6.
7.   ----------------------------
8.   deployment.apps/bar-deployment rolled back
```

這時 Deployment 已經回到了，沒有出問題的 revision=1 版本了。

```
1.   kubectl get all
2.   ----------------------------
3.   NAME                                   READY   STATUS    RESTARTS  AGE
4.   pod/bar-deployment-75bcfbd655-5b9z5    1/1     Running   0         17m
5.   pod/bar-deployment-75bcfbd655-6whzr    1/1     Running   0         17m
6.   pod/bar-deployment-75bcfbd655-zk88d    1/1     Running   0         17m
7.   pod/foo-deployment-6bbf665b47-dhndq    1/1     Running   0         50m
8.   pod/foo-deployment-6bbf665b47-pnjfs    1/1     Running   0         50m
9.
```

```
10.  NAME                      TYPE        CLUSTER-IP  EXTERNAL-IP  PORT(S)   AGE
11.  service/kubernetes        ClusterIP   10.96.0.1   <none>       443/TCP   23d
12.
13.  NAME                              READY    UP-TO-DATE    AVAILABLE    AGE
14.  deployment.apps/bar-deployment    3/3      3             3            17m
15.  deployment.apps/foo-deployment    2/2      2             2            50m
16.
17.  NAME                                       DESIRED   CURRENT   READY  AGE
18.  replicaset.apps/bar-deployment-75bcfbd655  3         3         3      17m
19.  replicaset.apps/bar-deployment-864b65d8b6  0         0         0      16m
20.  replicaset.apps/foo-deployment-6bbf665b47  2         2         2      50m
```

筆者碎碎念

我們上面大致練習了幾個比較實用的方式，可以發現 Deployment 的設計非常的
彈性以及簡潔，並且讓我們能將 Pod 設定在一起，大大地減少設定檔的數量。而
Deployment 因為可以簡單的設定水平擴展資源限制與請求等操作，使得許多進階
觀念，如藍綠部署、金絲雀部署……等更有可能被一般的後端工程師實現（真是
謝天謝地）。

參考資料

- Kubernetes 教學系列 - 滾動更新就用 Deployment
  https://blog.kennycoder.io/2021/01/09/Kubernetes%E6%95%99%E5%AD%
  B8%E7%B3%BB%E5%88%97-%E6%BB%BE%E5%8B%95%E6%9B%B4%E6%9
  6%B0%E5%B0%B1%E7%94%A8Deployment/

- 雲原生社區 -Deployment
  https://jimmysong.io/kubernetes-handbook/concepts/deployment.html

- Kubernetes Documentation-Deployment
  https://kubernetes.io/docs/concepts/workloads/controllers/
  deployment/#rolling-back-a-deployment

- [Kubernetes] Deployment Overview
  https://godleon.github.io/blog/Kubernetes/k8s-Deployment-Overview/

- Kubernetes 基礎教學（二）實作範例：Pod、Service、Deployment、Ingress
  https://cwhu.medium.com/kubernetes-implement-ingress-deployment-
  tutorial-7431c5f96c3e

Kubernetes — 實戰
做一個 StatefulSet

我們在上一章介紹了 Kubernetes Deployment，它是為了管理無狀態應用（Stateless）的資源物件。然而在實際的應用場景中，我們可能會遇到需要管理有狀態應用（Stateful）的服務，這就是我們本章要討論的主角：StatefulSet。

在學習 Statefulset 的過程中，我們可以慢慢體會到為什麼官方總是強調要盡可能將應用設計成 Stateless 而不是 Stateful，那是因為我們需要花費更多心力去維持應用內 Stateful 狀態，而不是可以輕易快速地重啟、刪除容器來恢復服務，所以我們需要先探討 StatefulSet 對於其保持 Stateful 的精神，才能幫助我們更進一步地活用。

8.1 StatefulSet 是什麼？

StatefulSet 是 Kubernetes 用於管理有狀態應用的工作負載物件。與 Deployment 和 ReplicaSet 不同，StatefulSet 中的每個 Pod 都有一個固定且唯一的名稱，並且在重新排程或重啟時，這個名稱將會保持不變。

StatefulSet 特別適合於需要穩定網路識別子和穩定儲存的應用，比如分散式系統，這些系統需要確保資料一致性和服務的高可用。當一個 StatefulSet 的 Pod 被重新建立，它可以重新掛載到之前的 PersistentVolumeClaim，確保資料的持久性。

StatefulSet 還提供了一種有序的部署和縮放機制。你可以定義 Pod 建立和刪除的順序，確保在建立或刪除多個 Pod 時，服務的可用性和資料的一致性得到保障。例如，你可以先建立所有的後端 Pod，然後再建立前端 Pod，或者在刪除時，先刪除前端 Pod，再刪除後端 Pod。

根據官方文件對其的敘述中點出，StatefulSet 對於需要滿足以下一個或多個需求的應用程序很有價值：

- 穩定的、唯一的網路識別子。
- 穩定的、持久的儲存。
- 有序的、優雅的部署和擴縮。
- 有序的、自動的滾動更新。

這些性質源自 Kubernetes 對於 StatefulSet 相關的限制：

- 給定 Pod 的儲存必須由 PersistentVolume Provisioner 基於所請求的 storage class 來製備，或者由管理員預先製備。
- 刪除或者擴縮 StatefulSet 並不會刪除它關聯的儲存卷。這樣做是為了保證資料安全，它通常比自動清除 StatefulSet 所有相關的資源更有價值。
- StatefulSet 當前需要無頭服務（Headless Service）來負責 Pod 的網路標識。
- 當刪除一個 StatefulSet 時，該 StatefulSet 不提供任何終止 Pod 的保證。為了實現 StatefulSet 中的 Pod 可以有序且體面地終止，可以在刪除之前將 StatefulSet 收縮到 0。
- 在預設 Pod 管理策略（OrderedReady）時使用滾動更新，可能進入需要人工干預才能修復的損壞狀態。

接下來我們將會以實戰演練來介紹這些 Kubernetes 在 StatefulSet 替我們實現的特性。

8.2 StatefulSet 中的有序命名及網路 ID

對於具有 N 個副本的 StatefulSet，該 StatefulSet 中的每個 Pod 將被分配一個整數序號，該序號在此 StatefulSet 上是唯一的。

預設情況下，這些 Pod 將被分配從 0 到 N-1 的序號，每個 Pod 的主機名根據 StatefulSet 的名稱和 Pod 的序號而衍生。主機名的組合格式為 $(statefulset)-$(index)。例如，一個包含三個 Pod 的 StatefulSet 會建立出名為 web-0、web-1、web-2 的 Pod。這些 Pod 可以使用無頭服務（Headless Service）來管理其網路領域。服務的格式為 $(serviceName).$(namespace).svc.cluster.local，其中 cluster.local 是叢集的域名。當每個 Pod 建立成功後，它們會得到對應的 DNS 子域，格式為 $(pod name).$(dns)。其中，所屬服務由 StatefulSet 的 serviceName 域設定。

▲ 圖 8-1 StatefulSet Ordinal Name

▶ 8.3 StatefulSet 中的穩定儲存

在 StatefulSet 中，對於每個定義的 VolumeClaimTemplate，每個 Pod 都會接收到一個 PersistentVolumeClaim。在下列的 nginx 範例中，每個 Pod 將會獲得一個基於 StorageClass 所建立的 1 Gib 的 PersistentVolume。若未明確宣告 StorageClass，則會使用預設的 StorageClass。當一個 Pod 被調度到某個節點時，它的 volumeMounts 將會掛載與其 PersistentVolumeClaims 關聯的 PersistentVolume。

NOTE

請注意，當 Pod 或 StatefulSet 被刪除時，關聯的 PersistentVolume 並不會被一併刪除。要刪除這些 PersistentVolume，必須透過手動方式來執行。

8.4 StatefulSet 中的 Headless Services

當我們不需要或不希望使用負載均衡和單獨的 Service IP 時，可以透過顯式指定 Cluster IP（spec.clusterIP）的值為「None」來建立 HeadlessService。我們可以使用一個 Headless Service 與其他服務發現機制進行接口，而不必與 Kubernetes 的實現捆綁在一起。Headless Services 不會獲得 Cluster IP，kube-proxy 不會處理這類服務，而且平台也不會為它們提供負載均衡或路由。

在 Kubernetes 中，serviceName 在 StatefulSet 設定中扮演重要角色。這個欄位指定了與 StatefulSet 關聯的 Headless Service 的名稱。這個 Headless Service 的主要目的是用於控制網路。傳統的 Service（非 Headless）會為背後的 Pod 提供一個單一的訪問點，並利用負載均衡來分配流量。但對於有狀態的應用來說，這可能並不是我們所需要的，因為我們可能希望能直接訪問某一個特定的 Pod，這就是 Headless Service 的作用。

當我們建立一個 Headless Service（.spec.clusterIP: None）並將其名稱指定給 StatefulSet 的 serviceName 屬性，Kubernetes 會為該 StatefulSet 中的每一個 Pod 建立一個獨立的 DNS 名稱。這些 DNS 名稱形式為 pod-name.service-name.namespace.svc.cluster.local，可以讓你直接訪問每一個 Pod，而不需要經過負載均衡。

所以我們可以如圖 8-2 架構實現對指定 Pod 溝通（見 8-6 頁）。

此外，使用 Headless Service，可以實現基於每個 Pod 的穩定網路識別碼，這對於分散式系統的一些要求（例如資料庫資料的一致性、領導者選舉等）是非常重要的。這也是為何在 StatefulSet 中需要指定 serviceName 的主要原因。

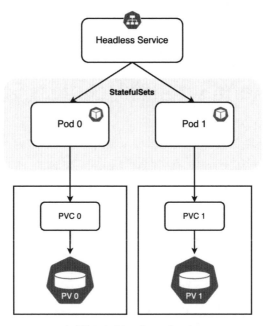

▲ 圖 8-2 Headless Service

8.5 StatefulSet 中的部署及擴縮保證

官方對於此 StatefulSet 的服務更新及穩定度，提出下列幾個原則：

● 對於包含 N 個副本的 StatefulSet，當部署 Pod 時，它們是依次建立的，順序為 0⋯⋯N-1。

● 當刪除 Pod 時，它們是逆序終止的，順序為 N-1⋯⋯0。

- 在將擴縮操作應用到 Pod 之前，所有 Pod 必須是 Running 和 Ready 狀態。
- 在一個 Pod 終止之前，所有的繼任者必須完全關閉。

在這幾個原則中，我們看得出為了保持 Stateful 應用的穩定，Kubernetes 對其的服務操作需要非常保守。在擴縮時，需要依序增長、依序縮減，還需要保證前一個服務順利執行並且完全關閉後才能對下一個 Pod 做更新。

⯈ 8.6 StatefulSet 中的更新策略

StatefulSet 的 .spec.updateStrategy 欄位讓你能夠設定並啟用或停用自動滾動更新的功能，這包含了對於 Pod 的容器、標籤、資源請求或限制，以及註解的更新。其提供兩種可選的值：

- OnDelete：當 StatefulSet 的 .spec.updateStrategy.type 設置為 OnDelete 時，其控制器將不會自動對 StatefulSet 中的 Pod 進行更新。如果使用者希望控制器針對 StatefulSet 的 .spec.template 的變更產生回應，建立新的 Pod，使用者必須手動刪除舊的 Pod。
- RollingUpdate：RollingUpdate 更新策略會自動對 StatefulSet 中的 Pod 進行滾動更新，這是 StatefulSet 預設的更新策略。

 NOTE

儘管 StatefulSet 提供了 RollingUpdate 策略進行自動的滾動更新，但由於每個 Pod 在 StatefulSet 中都擁有固定且獨特的識別，更新過程可能會相對複雜。不同於 Deployment，在 StatefulSet 的更新過程中，如果出現問題，可能會需要更多的人工介入來進行修復。

⏩ 8.7 實戰演練

以下為我們將會使用到 StatefulSet 搭配 Headless Service 的設定檔：

範例檔 **headless-service.yaml**

```
1.  # headless-service.yaml
2.  apiVersion: v1
3.  kind: Service
4.  metadata:
5.    name: foo-service
6.    labels:
7.      app: foo
8.  spec:
9.    ports:
10.    - port: 80
11.      name: http
12.    clusterIP: None
13.    selector:
14.      app: foo
```

重要參數：

- spec.clusterIP：設定為「None」表示這是一個 Headless Service，也就是說這個 Service 不會有一個代表其的 cluster IP，而是直接回傳後端 Pod 的 IP。
- spec.selector.app：這個欄位定義了一個標籤選擇器，用來選擇哪些 Pod 會被這個 Service 管理。在這裡，任何標籤為「app=nginx」的 Pod 都會被這個 Service 管理，包括我們後面準備產生的 StatefulSet。

範例檔 **statefulset.yaml**

```
1.  # statefulset.yaml
2.  apiVersion: apps/v1
```

```
3.    kind: StatefulSet
4.    metadata:
5.      name: foo-statefulset
6.    spec:
7.      updateStrategy:
8.        type: RollingUpdate
9.      selector:
10.       matchLabels:
11.         app: foo
12.     serviceName: "foo-service"
13.     replicas: 3
14.     template:
15.       metadata:
16.         labels:
17.           app: foo
18.       spec:
19.         terminationGracePeriodSeconds: 10
20.         containers:
21.           - name: foo
22.             image: nginx
23.             ports:
24.               - containerPort: 80
25.                 name: http
26.             volumeMounts:
27.               - name: pvc
28.                 mountPath: /data
29.     volumeClaimTemplates:
30.       - metadata:
31.           name: pvc
32.         spec:
33.           accessModes: [ "ReadWriteOnce" ]
34.           storageClassName: "hostpath"
```

```
35.        resources:
36.          requests:
37.            storage: 1Gi
```

重要參數：

- kind：表示我們要建立的 Kubernetes 物件的種類，在這裡是 StatefulSet。
- spec.serviceName：這個欄位指定了與 StatefulSet 相關聯的 Headless Service 的名稱。
- spec.updateStrategy：RollingUpdate 為預設的滾動更新，而另一個選項 OnDelete 則會停用自動更新，只有在手動刪除某個 Pod 之後，StatefulSet 控制器才會建立新的 Pod 來響應 .spec.template 的變更。
- spec.template.spec：
  - terminationGracePeriodSeconds：這個欄位指定了在 Kubernetes 發送終止信號到 Pod 中的所有容器後，應給予多長的緩衝期以供 Pod 內的容器進行清理工作。
- spec.volumeClaimTemplates.spec：這是一個 PersistentVolumeClaim 的模板列表，定義了 StatefulSet 中每個 Pod 所使用的 PersistentVolumeClaim。
  - accessModes：定義了可以如何存取 PersistentVolume。
  - storageClassName：定義了用於建立 PersistentVolume 的 StorageClass 的名稱。
  - resources.requests.storage：定義了 PersistentVolume 的大小需求。

接下來我們就來建立上面的設定檔：

```
1.  kubectl apply -f statefulset.yaml,headless-service.yaml
2.  ---------
3.  statefulset.apps/foo-statefulset created
4.  service/foo-service created
```

成功建立了 StatefulSet 以及對應 Headless Service。

查看 StatefulSet 的 Pod 名稱：

```
1. kubectl get pod
2. ------------
3. NAME                 READY   STATUS    RESTARTS   AGE
4. foo-statefulset-0    1/1     Running   0          6m31s
5. foo-statefulset-1    1/1     Running   0          6m28s
6. foo-statefulset-2    1/1     Running   0          6m25s
```

在此之前我們也可以藉由 -w（watch）參數，來驗證 StatefulSet 是否如我們預料中的依序增長：

```
1.  kubectl get pod -w
2.  ------------
3.  NAME                 READY   STATUS             RESTARTS   AGE
4.  foo-statefulset-0    0/1     Pending            0          0s
5.  foo-statefulset-0    0/1     Pending            0          0s
6.  foo-statefulset-0    0/1     ContainerCreating  0          0s
7.  foo-statefulset-0    1/1     Running            0          3s
8.  foo-statefulset-1    0/1     Pending            0          0s
9.  foo-statefulset-1    0/1     Pending            0          0s
10. foo-statefulset-1    0/1     ContainerCreating  0          0s
11. foo-statefulset-1    1/1     Running            0          3s
12. foo-statefulset-2    0/1     Pending            0          0s
13. foo-statefulset-2    0/1     Pending            0          0s
14. foo-statefulset-2    0/1     ContainerCreating  0          0s
15. foo-statefulset-2    1/1     Running            0          3s
```

別忘了在同一時間 StatefulSet 掛載的 PVC 也會依序建立：

```
1. kubectl get pvc -w
2. ------------
3. NAME            STATUS   VOLUME   CAPACITY   ACCESS MODES
   STORAGECLASS    AGE
```

```
4.  pvc-foo-statefulset-0    Pending
    hostpath        0s
5.  pvc-foo-statefulset-0    Pending
    hostpath        0s
6.  pvc-foo-statefulset-0    Pending    pvc-32fa04cd-b0fb-41aa-808e-
    a756fa34ac94    0                   hostpath        0s
7.  pvc-foo-statefulset-0    Bound      pvc-32fa04cd-b0fb-41aa-808e-
    a756fa34ac94    1Gi        RWO      hostpath        0s
8.  pvc-foo-statefulset-1    Pending
    hostpath        0s
9.  pvc-foo-statefulset-1    Pending
    hostpath        0s
10. pvc-foo-statefulset-1    Pending    pvc-b57b84f2-d3c6-4afd-83a4-
    3011aa53f9a6    0                   hostpath        0s
11. pvc-foo-statefulset-1    Bound      pvc-b57b84f2-d3c6-4afd-83a4-
    3011aa53f9a6    1Gi        RWO      hostpath        0s
12. pvc-foo-statefulset-2    Pending
    hostpath        0s
13. pvc-foo-statefulset-2    Pending
    hostpath        0s
14. pvc-foo-statefulset-2    Pending    pvc-5be819cb-1afa-4b48-aa4a-
    0dfdfb99249f    0                   hostpath        0s
15. pvc-foo-statefulset-2    Bound      pvc-5be819cb-1afa-4b48-aa4a-
    0dfdfb99249f    1Gi        RWO      hostpath        0s
```

為了驗證 Headless Service 是否已經正確設定並運行，我們可以透過以下步驟執行檢查。

首先，我們可以使用 kubectl exec 指令進入一個具體的 Pod。在這個例子中，我們選擇進入「foo-statefulset-0」Pod，確保我們可以在和 Headless Service 相同的內網環境中進行測試。這可以透過以下指令達成：

```
1.  kubectl exec -it foo-statefulset-0 -- bash
```

接著，我們在「foo-statefulset-0」Pod 內部，使用 curl 指令嘗試透過 Headless Service 來訪問「foo-statefulset-1」Pod。因為我們已經在 Headless Service 的網路環境中，我們可以直接使用 Service 名稱「foo-service」來進行訪問。該指令如下：

```
1.  curl foo-statefulset-1.foo-service
2.  ------------
3.  <!DOCTYPE html>
4.  <html>
5.  <head>
6.  <title>Welcome to nginx!</title>
7.  <style>
8.  html { color-scheme: light dark; }
9.  body { width: 35em; margin: 0 auto;
10. font-family: Tahoma, Verdana, Arial, sans-serif; }
11. </style>
12. </head>
13. <body>
14. <h1>Welcome to nginx!</h1>
15. <p>If you see this page, the nginx web server is successfully
    installed and
16. working. Further configuration is required.</p>
17.
18. <p>For online documentation and support please refer to
19. <a href="http://nginx.org/">nginx.org</a>.<br/>
20. Commercial support is available at
21. <a href="http://nginx.com/">nginx.com</a>.</p>
22.
23. <p><em>Thank you for using nginx.</em></p>
24. </body>
25. </html>
```

我們成功地從「foo-statefulset-0」Pod 訪問到了「foo-statefulset-1」Pod，並且得到了 Nginx 歡迎頁面的回應。這就證明了我們的 Headless Service 已經成功建立並運行。

總結來說，這段操作完成了以下任務：

- 進入「foo-statefulset-0」Pod，使我們可以在和 Headless Service 同一個網路環境中進行檢測。
- 使用 Headless Service 訪問另一個 Pod「foo-statefulset-1」，並成功獲得回應。

透過這個流程，我們可以清楚地驗證 Headless Service 是否已經正確設定並運行。

接下來也可以對 StatefulSet 的更新行為進行觀察：

```
1.   kubectl rollout restart statefulset foo-statefulset
2.   ------------
3.   statefulset.apps/foo-statefulset restarted
```

在我們執行刪除同時也可以使用 -w（watch）參數來觀察 Pod 的變化：

```
1.   kubectl get pod -w
2.   ------------
3.   NAME                READY   STATUS        RESTARTS   AGE
4.   foo-statefulset-0   1/1     Running       0          7s
5.   foo-statefulset-1   1/1     Running       0          11s
6.   foo-statefulset-2   1/1     Running       0          15s
7.   foo-statefulset-2   1/1     Terminating   0          21s
8.   foo-statefulset-2   0/1     Terminating   0          21s
9.   foo-statefulset-2   0/1     Terminating   0          21s
10.  foo-statefulset-2   0/1     Terminating   0          21s
11.  foo-statefulset-2   0/1     Pending       0          0s
12.  foo-statefulset-2   0/1     Pending       0          0s
```

```
13. foo-statefulset-2   0/1   ContainerCreating   0        0s
14. foo-statefulset-2   1/1   Running             0        3s
15. foo-statefulset-1   1/1   Terminating         0        20s
16. foo-statefulset-1   0/1   Terminating         0        21s
17. foo-statefulset-1   0/1   Terminating         0        21s
18. foo-statefulset-1   0/1   Terminating         0        21s
19. foo-statefulset-1   0/1   Pending             0        0s
20. foo-statefulset-1   0/1   Pending             0        0s
21. foo-statefulset-1   0/1   ContainerCreating   0        0s
22. foo-statefulset-1   1/1   Running             0        3s
23. foo-statefulset-0   1/1   Terminating         0        20s
24. foo-statefulset-0   0/1   Terminating         0        21s
25. foo-statefulset-0   0/1   Terminating         0        21s
26. foo-statefulset-0   0/1   Terminating         0        21s
27. foo-statefulset-0   0/1   Pending             0        0s
28. foo-statefulset-0   0/1   Pending             0        0s
29. foo-statefulset-0   0/1   ContainerCreating   0        0s
30. foo-statefulset-0   1/1   Running             0        3s
```

可以發現 StatefulSet 照著預期中的，依序由最後重新啟動並且等到每個 Pod 的前一位依賴者狀態為 Running 後才開始更新。

▷ 8.8 刪除 StatefulSet

我們可以像刪除 Kubernetes 中的其他資源一樣刪除 StatefulSet：使用 kubectl delete 指令，並按照檔案或者名字指定 StatefulSet。

但需要注意的是刪除 StatefulSet 管理的 Pod 並不會刪除關聯的 PVC。這是為了確保你有機會在刪除卷之前從卷中複製資料。在 Pod 已經終止後刪除 PVC 可能會觸發刪除背後的 PV，具體取決於儲存類型和回收策略。

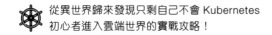

刪除剛剛建立的 foo-statefulset：

```
1.  kubectl delete statefulset foo-statefulset
2.  -----------
3.  statefulset.apps "foo-statefulset" deleted
```

同時我們也使用 -w（watch）觀察 StatefulSet 被刪除時的操作：

```
1.  kubectl get pod -w
2.
3.  -----------
4.  NAME                READY   STATUS        RESTARTS   AGE
5.  foo-statefulset-0   1/1     Running       0          10m
6.  foo-statefulset-1   1/1     Running       0          10m
7.  foo-statefulset-2   1/1     Running       0          10m
8.  foo-statefulset-1   1/1     Terminating   0          10m
9.  foo-statefulset-0   1/1     Terminating   0          10m
10. foo-statefulset-2   1/1     Terminating   0          11m
11. foo-statefulset-2   0/1     Terminating   0          11m
12. foo-statefulset-2   0/1     Terminating   0          11m
13. foo-statefulset-2   0/1     Terminating   0          11m
14. foo-statefulset-0   0/1     Terminating   0          10m
15. foo-statefulset-0   0/1     Terminating   0          10m
16. foo-statefulset-0   0/1     Terminating   0          10m
17. foo-statefulset-1   0/1     Terminating   0          11m
18. foo-statefulset-1   0/1     Terminating   0          11m
19. foo-statefulset-1   0/1     Terminating   0          11m
```

這時我們可以發現在刪除過程中，StatefulSet 將並行的刪除所有 Pod，在刪除一個 Pod 前不會等待它的順序後繼者終止。

而我們需要注意的是伴隨著 StatefulSet 動態產生的 PVC 並不會隨著該 StatefulSet 資源被刪除時一併移除：

```
1.  kubectl get pvc
2.  ------------
3.  NAME                    STATUS    VOLUME
    CAPACITY    ACCESS MODES    STORAGECLASS    AGE
4.  pvc-foo-statefulset-0    Bound    pvc-32fa04cd-b0fb-41aa-808e-
    a756fa34ac94    1Gi         RWO              hostpath        52m
5.  pvc-foo-statefulset-1    Bound    pvc-b57b84f2-d3c6-4afd-83a4-
    3011aa53f9a6    1Gi         RWO              hostpath        52m
6.  pvc-foo-statefulset-2    Bound    pvc-5be819cb-1afa-4b48-aa4a-
    0dfdfb99249f    1Gi         RWO              hostpath        52m
```

我們需要額外處理這些 StatefulSet 產生的儲存資源：

```
1.  kubectl delete pvc pvc-foo-statefulset-0 pvc-foo-statefulset-1
    pvc-foo-statefulset-2
2.  ------------
3.  persistentvolumeclaim "pvc-foo-statefulset-0" deleted
4.  persistentvolumeclaim "pvc-foo-statefulset-1" deleted
5.  persistentvolumeclaim "pvc-foo-statefulset-2" deleted
```

> 我們需要記得刪除實際場景中用到的 PVC 等持久化儲存介質，最理想的是基於
> 環境、儲存設定和製備方式，按照必需的步驟保證回收所有的儲存。

🗄 筆者碎碎念

瞭解並妥善運用 StatefulSet 是一項關鍵任務，特別是當我們的應用需要維護其
狀態並在分散式環境中提供一致性時。StatefulSet 不僅僅是一種管理 Pod 的方
式，它也提供了重要的特性，例如穩定的網路識別和持久儲存，這也意味著我們
需要更深入的理解和掌握更多的知識，包括 Pod、Service、Persistent Volume、
DNS 解析等 Kubernetes 的核心概念。與 Deployment 一樣，我們也需要注意到
StatefulSet 的更新策略，以及如何有效地管理和維護在 StatefulSet 中運行的每個
Pod。

參考資料

- Kubernetes StatefulSet
  https://kubernetes.io/zh-cn/docs/concepts/workloads/controllers/statefulset/

- 刪除 StatefulSet
  https://kubernetes.io/zh-cn/docs/tasks/run-application/delete-stateful-set/

- 強制刪除 StatefulSet 中的 Pod
  https://kubernetes.io/zh-cn/docs/tasks/run-application/force-delete-stateful-set-pod/

- StatefulSet 基礎
  https://kubernetes.io/zh-cn/docs/tutorials/stateful-application/basic-stateful-set/#deleting-statefulsets

- Kubernetes Service
  https://kubernetes.io/zh-cn/docs/concepts/services-networking/service/#headless-services

Part 4
進階基礎篇

我就知道事情
沒有那麼單純

我們經常在學習新事物時，可能因為一開始跟著教學快速獲得的收穫而自信滿滿，卻不知不覺經歷達克效應中愚昧山丘的階段，接下來就讓我們來揭開藏在魔鬼中的細節吧。

CHAPTER

09

Kubernetes — Kustomize 是什麼？

經歷了各種實戰演練後，我們已經可以簡單聲明並部署 Kubernetes 服務，但別忘了在實際工作環境中的專案管理複雜度總是遠大於我們在單純的本地環境練習，這時我們可能會遇到跨團隊開發、多環境設定、設定檔管理等棘手問題，而 Kubernetes 則為我們提供了 Kustomize 這個解決方案。

9.1 Kustomize 在 Kubernetes 中的定位

Kustomize 是由 Kubernetes SIGs（Special Interest Groups）專案社群負責，它們負責著 Kubernetes 專案中的特定部分。

SIGs 是一種對某一特定主題有共同興趣的群體組成的組織形式，在 Kubernetes 社群中扮演著極為重要的角色。每個 SIGs 都有一個特定的任務範圍或特定的專案，包括 Architecture、Networking、Storage、Auth 等多個領域，同時他們也貢獻了許多好用的工具與套件，像是 Kubespray、Kind、Krew 等，還有稍後會提到的 Kustomize。

在 Kubernetes v1.14 版本後，Kustomize 已經成為了 kubectl 的內建指令，可以簡單地輸入「kubectl apply -k ...」使用其相關功能。

9.2 Kustomize 介紹

Kustomize 是一種 Kubernetes 資源的設定管理工具。它的設計目標是簡化和標準化 Kubernetes 的設定管理流程。

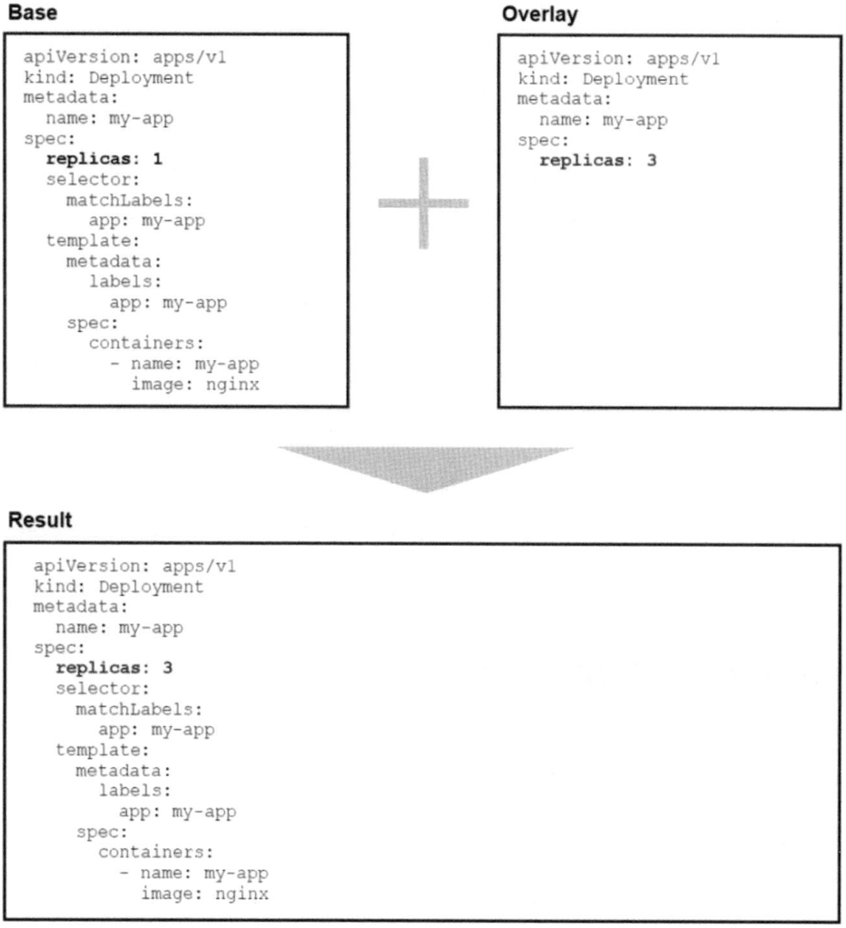

Base

```
apiVersion: apps/v1
kind: Deployment
metadata:
  name: my-app
spec:
  replicas: 1
  selector:
    matchLabels:
        app: my-app
  template:
    metadata:
      labels:
        app: my-app
      spec:
        containers:
          - name: my-app
            image: nginx
```

Overlay

```
apiVersion: apps/v1
kind: Deployment
metadata:
  name: my-app
spec:
  replicas: 3
```

Result

```
apiVersion: apps/v1
kind: Deployment
metadata:
  name: my-app
spec:
  replicas: 3
  selector:
    matchLabels:
        app: my-app
  template:
    metadata:
      labels:
        app: my-app
      spec:
        containers:
          - name: my-app
            image: nginx
```

▲ 圖 9-1 Kustomize

這裡列出 Kustomize 嘗試解決的幾個主要問題：

1. 模板的複雜性：在 Kustomize 出現之前，許多 Kubernetes 的設定管理工具
   會使用模板，例如 Helm。模板的方法雖然靈活，但也複雜且容易出錯。相
   反地，Kustomize 採取一種基於原始的 YAML 檔案的「generate-patch」方
   法，這種方法比模板更簡單、直觀。

2. 環境間的設定差異：在不同的環境中（例如開發、測試和生產環境），
   Kubernetes 的設定可能會有些微的差異。Kustomize 提供一種簡單的方式來
   管理這些差異，而不必為每個環境建立一套完整的設定。

3. 設定的重複和共享：在大型的 Kubernetes 專案中，可能有許多共享和重複
   的設定。Kustomize 允許用戶建立「基礎」設定，並根據需要在各個場景中
   進行客製化，這使得設定更加乾淨和可維護。

4. 策略和設定的分離：Kustomize 提供了一種將策略（例如 Kubernetes 物件
   的名稱和命名空間）和設定分離的方式。這讓管理和調整策略變得更加簡
   單，並且使得設定能更容易地在不同的團隊和專案之間共享。

Kustomize 提供了類似於 sed 功能，將各個設定值藉由 replace 的方式，改寫到
YAML 檔案中，達到同樣的基礎設定檔卻能夠在不同的環境中，產生不同的設定。

▶ 9.3 Kustomize 安裝

1. 使用 Homebrew 安裝 Kustomize（macOS）：

```
1.  brew install kustomize
```

2. 或直接使用 kubectl 內建的 Kustomize 指令：

```
1.  kubectl kustomize -h
```

▶ 9.4 基本指令

輸出 Kustomize 產生的結果：

```
1.  kubectl kustomize <kustomization_dir>
```

查看 Ksutomize 結果與線上差異：

```
1.  kubectl diff -k <kustomization_dir>
```

執行 Kustomize 產生的結果：

```
1.  kubectl apply -k <kustomization_dir>
```

刪除 Kustomize 產生的資源：

```
1.  kubectl delete -k <kustomization_dir>
```

▷ 9.5 實戰演練

接下來我們馬上使用 Kustimoize 簡化設定，體驗它如何幫助我們抽象出設定檔結構。

9.5.1 範例一：Simple Kustomization

API 服務資料夾結構：

```
foo
├── deployment.yaml
└── service.yaml
bar
├── deployment.yaml
└── service.yaml
```

假如我們現在有兩個極為耦合的 API 服務設定檔 - foo、bar，兩者的差異只在於 Image.Name 上，使用 Kustomization 可以避免出現高度重複的資料夾結構。

基本 Kustomization 資料夾結構：

```
.
├── deployment.yaml
├── kustomization.yaml
└── service.yaml
```

由上面最單純的 Kuberentes 設定檔資料夾結構，可以以一個服務為單位，使用 Kustomization 設定檔收斂、管理整個服務。

設定檔內容：

範例檔 **simple / deployment.yaml**

```
1.  # simple/deployment.yaml
2.  apiVersion: apps/v1
3.  kind: Deployment
4.  metadata:
5.    name: deployment
6.    labels:
7.      type: demo
8.  spec:
9.    replicas: 2
10.   selector:
11.     matchLabels:
12.       type: demo
13.   template:
14.     metadata:
15.       labels:
16.         type: demo
17.     spec:
18.       containers:
19.         - name: api-service
```

```
20.              image: mikehsu0618/api-service:tag
21.              ports:
22.                - containerPort : 8080
```

範例檔 **simple / service.yaml**

```
1.  # simple/service.yaml:
2.  apiVersion: v1
3.  kind: Service
4.  metadata:
5.    name: service
6.  spec:
7.    selector:
8.      type: demo
9.    type: LoadBalancer
10.   ports:
11.     - protocol: TCP
12.       port: 8000
13.       targetPort: 8080
```

這邊 deployment.yaml、service.yaml 就如我們前面使用的一樣，是一個簡單暴露 port 8000 的 API 服務，接下來我們可以在 kustomization.yaml 來抽象出設定檔中的變數。

範例檔 **simple / kustomization.yaml**

```
1.  # simple/kustomization.yaml
2.  apiVersion: kustomize.config.k8s.io/v1beta1
3.  kind: Kustomization
4.
5.  namePrefix: foo-
6.  nameSuffix: -v1
7.
8.  commonLabels:
```

```
9.     by: kustomization
10.
11.  commonAnnotations:
12.     note: Hello, I am foo!
13.
14.  images:
15.    - name: mikehsu0618/api-service
16.      newName: mikehsu0618/foo
17.      newTag: v1.0.0
18.
19.  resources:
20.    - deployment.yaml
21.    - service.yaml
```

此設定檔可以直觀地看出我們想從 foo、bar 這兩個服務設定檔中，解耦出 name、label、annotations、images 等相關資訊。

重要參數：

- resources：導入需要被 Kustomization 覆寫的設定檔路徑。
- namePrefix & nameSuffix：在對應資源名稱添加指定前後綴，用以區分服務名稱。
- commonLabels：在所有導入資源皆附上對應 Label，如果已經存在就覆蓋。
- commonAnnotations：在所有導入資源皆附上對應 Annotations，如果已經存在就覆蓋。
- images：使用 images.name 從導入設定檔中指定要覆寫的 newName、newTag。

如此一來，我們只要抽象出設定檔的變數，即可在 Kustomization.yaml 中輕鬆設定不同種服務了。

最後我們就來查看 Kustomize 為我們產生的結果：

```
1.  kubectl kustomize ./
2.  ------
3.  apiVersion: v1
4.  kind: Service
5.  metadata:
6.    annotations:
7.      note: Hello, I am foo!
8.    labels:
9.      by: kustomization
10.   name: foo-service-v1
11. spec:
12.   ports:
13.   - port: 8000
14.     protocol: TCP
15.     targetPort: 8080
16.   selector:
17.     by: kustomization
18.     type: demo
19.   type: LoadBalancer
20. ---
21. apiVersion: apps/v1
22. kind: Deployment
23. metadata:
24.   annotations:
25.     note: Hello, I am foo!
26.   labels:
27.     by: kustomization
28.     type: demo
29.   name: foo-deployment-v1
30. spec:
31.   replicas: 2
```

```
32.    selector:
33.      matchLabels:
34.        by: kustomization
35.        type: demo
36.    template:
37.      metadata:
38.        annotations:
39.          note: Hello, I am foo!
40.        labels:
41.          by: kustomization
42.          type: demo
43.      spec:
44.        containers:
45.        - image: mikehsu0618/foo: v1.0.0
46.          name: foo
47.          ports:
48.          - containerPort: 8080
```

順利將我們抽象出來的 API 服務設定檔，覆寫成 foo v1。

9.5.2 範例二：Overlay Kustomization

API 服務資料夾結構：

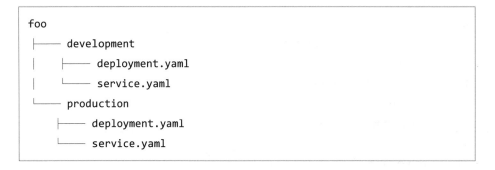

```
foo
├── development
│   ├── deployment.yaml
│   └── service.yaml
└── production
    ├── deployment.yaml
    └── service.yaml
```

假如我們有一個 foo 服務，但它在不同的環境中有不同的設定參數，在不依靠 Kustomize 及其他工具時，我們可能會產生的資料夾結構會如上所示。

```
├── base
│   ├── deployment.yaml
│   ├── kustomization.yaml
│   └── service.yaml
└── overlays
    ├── development
    │   └── kustomization.yaml
    └── production
        └── kustomization.yaml
```

現在我們使用與先前相同的操作。透過 Kustomize，我們可以抽象出設定檔模板之間的參數差異，藉此簡化多個環境下高度耦合的設定檔。使用 Kustomize 的強大功能，可以讓我們避免在實務情形中因為只需修改一個 env 參數，卻必須重複複製整份設定檔的尷尬情況。

設定檔內容：

範例檔 **overlay / overlays / development / kustomization.yaml**

```
1.  # overlays/development/kustomization.yaml
2.  resources:
3.    - ../../base
4.
5.  namePrefix: dev-
6.
7.  namespace: dev-namesapce
8.
9.  commonLabels:
10.    type: dev-demo
```

```
11.     app: dev-foo
12.
13.   commonAnnotations:
14.     note: Hello, I am development!
15.
16.   images:
17.     - name: mikehsu0618/api-service
18.       newTag: development
19.
20.   patches:
21.     - patch: |
22.         - op: replace
23.           path: /metadata/name
24.           value: the-dev-development
25.         - op: replace
26.           path: /spec/template/spec/containers/0/name
27.           value: the-dev-container
28.       target:
29.         group: apps
30.         kind: Deployment
31.         version: v1
32.         name: foo-deployment-v1
```

範例檔 **overlay / overlays / production / kustomization.yaml**

```
1.   # overlays/production/kustomization.yaml
2.   resources:
3.     - ../../base
4.
5.   namePrefix: prod-
6.
7.   namespace: production-namesapce
```

```
8.
9.  commonLabels:
10.    type: prod-demo
11.    app: prod-foo
12.
13.  commonAnnotations:
14.    note: Hello, I am Production!
15.
16.  images:
17.    - name: mikehsu0618/api-service
18.      newTag: production
19.
20.  patches:
21.    - patch: |
22.        - op: replace
23.          path: /metadata/name
24.          value: the-prod-development
25.        - op: replace
26.          path: /spec/template/spec/containers/0/name
27.          value: the-prod-container
28.      target:
29.        group: apps
30.        kind: Deployment
31.        version: v1
32.        name: foo-deployment-v1
```

從 overlays 中的設定檔可以看出，我們只將環境變數抽出來單獨替換，並且同樣地保持了高內聚低耦合的精神。

重要參數：

- resources：在此處的 resources 指向另一個 kustomization.yaml，並且疊加設定上去。
- namePrefix：可以建立在 resource name 原本的基礎上繼續添加前後綴。
- patches：此參數可以用 path、value 指定符合 target 條件資源中的任何 key value，並且使用各種 op 達到替換、添加或移除等十分彈性靈活的操作。
- others：在每個最頂層的設定中，kustomization.yaml 皆可以實現疊加，即為這種 CreateOrUpdate 的概念，使得 namespace、commonLabels、commonAnnotations、images 可以脫離對其他設定檔的依賴。

接下來我們針對 deployment.yaml 分別來看看套用了不同環境變數的結果：

1. Development：

```
1.   kubectl kustomize ./overlays/development
2.   ---------------------------------------
3.   ...
4.   ---
5.   apiVersion: apps/v1
6.   kind: Deployment
7.   metadata:
8.     annotations:
9.       note: Hello, I am development!
10.    labels:
11.      app: dev-foo
12.      by: kustomization
13.      org: Corporation
14.      type: dev-demo
15.    name: dev-the-dev-development
16.  spec:
17.    replicas: 2
```

```
18.    selector:
19.      matchLabels:
20.        app: dev-foo
21.        by: kustomization
22.        org: Corporation
23.        type: dev-demo
24.    template:
25.      metadata:
26.        annotations:
27.          note: Hello, I am development!
28.        labels:
29.          app: dev-foo
30.          by: kustomization
31.          org: Corporation
32.          type: dev-demo
33.      spec:
34.        containers:
35.        - image: mikehsu0618/foo: v1.0.0
36.          name: the-dev-container
37.          ports:
38.          - containerPort: 8080
```

2. Production：

```
1.  kubectl kustomize ./overlays/production
2.  ----------------------------------------
3.  ...
4.  ---
5.  apiVersion: apps/v1
6.  kind: Deployment
7.  metadata:
8.    annotations:
```

```
9.       note: Hello, I am Production!
10.    labels:
11.      app: prod-foo
12.      by: kustomization
13.      type: prod-demo
14.    name: prod-the-prod-development
15.    namespace: production-namesapce
16.  spec:
17.    replicas: 2
18.    selector:
19.      matchLabels:
20.        app: prod-foo
21.        by: kustomization
22.        type: prod-demo
23.    template:
24.      metadata:
25.        annotations:
26.          note: Hello, I am Production!
27.        labels:
28.          app: prod-foo
29.          by: kustomization
30.          type: prod-demo
31.      spec:
32.        containers:
33.        - image: mikehsu0618/foo: v1.0.0
34.          name: the-prod-container
35.          ports:
36.          - containerPort: 8080
```

可以看到 Kustomization 輕鬆為我們解決多環境部署時，造成的設定檔重複冗
餘的問題。

▶ 9.6 Kustomize 進階功能

藉由前面的兩個範例，可以深深體會到 Kustomize 的便利性，而為了涵蓋到我們實務上更細膩的設定，Kubernetes 官方也提出了一些直覺方便的參數來解決設定檔管理問題。

9.6.1 Patches

在 patches 設定中，可以簡單理解為「選定物件，然後對指令欄位進行修改」。

- target：可以透過 group、version、kind、name、namespace、標籤選擇器和註釋選擇器來選擇資源，選擇一個或多個匹配所有指定欄位的資源來應用 patch。
- patch：可以使用 strategic merge patch 或 Json6902 patch，也可以是一個 patch 檔案或者是 inline string，同時可以針對一個或一個以上的資源分別做設定。
- patchesStrategicMerge：此方法可以使用導入 patch 檔案並使用 resource name 來做匹配，將設定套用在指定資源上。
- patchesJson6902：JSON Patch（Json6902）是一個在 RFC 6902 中定義的標準，用於描述 JSON 檔案中的更改，而 Kustomize 使用其巧妙的替換設定檔中的特定設定。
- add：這個操作會在給定的路徑添加一個值。如果路徑指向的位置已經存在一個值，add 會代替那個值。如果路徑不存在，add 會建立它。
- remove：這個操作會移除指定路徑的值。
- replace：這個操作會替換給指定路徑的現值。如果路徑不存在，替換操作會失敗。

以下為 patchStrategicMerge 與 patchJson6902 的相關設定範例：

```
1.   patchesJson6902:
2.   - target:
3.       version: v1
4.       kind: Deployment
5.       name: my-deployment
6.      path: add_init_container.yaml
7.   - target:
8.       version: v1
9.       kind: Deployment
10.       name: my-deployment
11.     patch: |-
12.       - op: add
13.         path: /some/new/path
14.         value: value
15.       - op: replace
16.         path: /some/existing/path
17.         value: "new value"
18.
19.   patchesStrategicMerge:
20.   - deployment_increase_replicas.yaml
21.   - deployment_increase_memory.yaml
22.   - |-
23.     apiVersion: apps/v1
24.     kind: Deployment
25.     metadata:
26.       name: nginx
27.     spec:
28.       template:
29.         spec:
30.           containers:
31.             - name: nginx
32.               image: nignx: latest
```

> 原本與 Patches 平行層級的 PatchesStrategicMerge 和 patchesJson 6902 設定，在 Kustomize v5.0.0 中，已經被整合進 Patches.patch 中並準備被棄用。

9.6.2 ConfigGenerator

當需要區分不同的變數或環境設定時，ConfigMap 以及 Secret 資源也佔有舉足輕重的地位，Kustomize 使用 ConfigGenerator、SecretGenerator 讓我們管理設定，且避免寫出冗餘的設定檔。

- name：產生列表中該命名的 ConfigMap。
- behavior：它允許我們確定如何處理已存在的 ConfigMap 資源，無論是建立新的、替換既有的，還是與既有的進行合併。透過選擇合適的行為模式，可以更有效地管理 ConfigMap，避免重複或不一致的設定。
- files：以檔案為單位對 ConfigMap 進行設定，可以將導入的檔案在 ConfigMap 中重新命名。
- literals：以 key / value 對 ConfigMap 進行設定。
- env：允許從環境變數檔案中生成 ConfigMap。環境變數檔案是一種特殊類型的檔案，其中包含了一組環境變數，每行一個，格式為 KEY=VALUE。

以下為 ConfigGenerator 的相關設定範例：

```
1.  configMapGenerator:
2.  - name: my-java-server-props
3.    behavior: merge
4.    files:
5.    - application.properties
6.  - name: my-java-server-env-file-vars
7.    envs:
8.    - my-server-env.properties
9.  - name: my-java-server-env-vars
```

```
10.    literals:
11.    - JAVA_HOME=/opt/java/jdk
12.  - name: dashboards
13.    files:
14.    - mydashboard.json
15.  - name: app-whatever
16.    files:
17.    - myFileName.ini=whatever.ini
```

9.6.3 SecretGenerator

SecretGenerator 與 ConfigGenerator 參數設定相當類似，列表中的每個項目都
可以產生出一個 Secret。

- type：指定的 Secret 資源 type，例如：Opaque、kuberentes.io/tls 等等。

以下為 SecretGenerator 的相關設定範例：

```
1.  secretGenerator:
2.  - name: app-tls
3.    files:
4.    - secret/tls.cert
5.    - secret/tls.key
6.    type: "kubernetes.io/tls"
7.  - name: env_file_secret
8.    envs:
9.    - env.txt
10.    type: Opaque
11.  - name: secret-with-annotation
12.    files:
13.    - app-config.yaml
14.    type: Opaque
```

9.6.4 HelmChart

相信透過前面的介紹，很多人在管理設定檔時會聯想到 Helm。兩個較為不同的是，Helm 提供文件模板（Go template）通過部署時渲染成設定檔，是一種相較有侵入性的做法，而 Kustomize 使用原生 Kuberenetes 資源物件，無須模板參數化或侵入，並且使用 patch 來生成最終的設定檔。在定位方面，Helm 專注於應用的複雜性及生命週期管理（install、upgrade、rollback），而 Kustomize 則相較輕量、透過 Kubernetes 宣告式 API 設定檔管理。

- helmGlobals：kustomize 會到下面 chartHome 宣告的 Chart Repo 下載 Chart 放到這個資料夾。
- repo：Chart Repo 位置。
- version：Chart 版本。
- releaseName：該應用的 release 名稱。
- valuesFile：為 helm install 時套用的 vlaues.yaml 檔案位置，additionalValuesFiles 則可以繼續添加其他 vlaues.yaml。

以下為 HelmChart 的相關設定範例：

```
1.   helmGlobals:
2.     chartHome: ./charts
3.   helmCharts:
4.   - name: minecraft
5.     repo: https://kubernetes-charts.storage.googleapis.com
6.     version: v1.2.0
7.     releaseName: test
8.     namespace: testNamespace
9.     valuesFile: values.yaml
10.    additionalValuesFiles:
11.    - values-file-1.yaml
```

從異世界歸來發現只剩自己不會 Kubernetes
初心者進入雲端世界的實戰攻略！

Kubernetes —
路由守護神 Ingress

在前面章節中，我們介紹了 Service 這個元件，並且說明了如何利用它讓叢集中的 Pod 可以被外部的用戶或系統存取。然而，每個 Service 都需要指定一個對外的 port number，並且在 Node 上進行 port mapping。換句話說，每個 Service 都需要在主機上佔用一個特定的連接埠，用於與外部世界進行通訊。這就代表，如果我們有大量的 Service，就需要管理大量的 port number，這將使系統管理變得相當複雜。

再者，我們的網站和應用程式通常都會使用標準的 HTTP（port 80）或 HTTPS（port 443）連接埠。如果我們需要在 URL 中加入特定的連接埠號，這將大大降低網站的易用性，因為使用者需要記住並輸入這些特殊的連接埠號。因此，為了提高實用性和易用性，我們需要尋找一種方法，可以讓我們的 Service 在不需要指定特殊連接埠號的情況下被外部訪問。

10.1 Ingress 是什麼？

Ingress 是 Kubernetes 中的一種資源類型，主要用於管理外部訪問叢集中的服務路由。它可以幫助我們統一管理對外的 port number，並且根據訪問的 hostname 或是 pathname，將請求轉發到相對應的 Service。簡單來說，你可以想象它就像一個郵差，根據位址來決定應該將郵件投遞到哪裡。

此外，Ingress 還可以扮演一個上層的 LoadBalancer 的角色。LoadBalancer 的主要作用是分散和平衡向我們的服務發送的流量，以確保服務的穩定性和可靠性。

另外，使用 Kubernetes Ingress 還有一個優點，那就是它會統一打開 http 的 80 port 以及 https 的 443 port。這解決了我們在管理眾多服務時可能遇到的 port number 紊亂的問題，使我們可以更方便地管理我們的服務。

舉個例子，假設我們有一個應用，其中包括前端和後端兩個部分，並分別運行在不同的服務中。我們可以設置 Ingress，讓來自外部的請求在訪問 http://our-app.com/ 時，被導向到前端的服務，而當訪問 http://our-app.com/api 時，則被導向到後端的服務。這樣，我們就可以使用一個單一的進入點來管理我們的應用，並根據路徑來路由請求。

10.2 Ingress 的作用

Ingress 負責的事情主要被定義為下面幾項：

- 將不同路徑的請求對應到各自的 Service（Give Services Externally-Reachable URLs）：只要透過設定好的 hostname 跟 pathname 就可以觸及到對應的 Services 進而存取其對應的 Pods。
- 流量的負載均衡（Load Balance Traffic）：例如負載均衡算法、後端權重方案等。
- 支援 SSL Termination：支援 https 的傳輸層安全協定並且擔任起解密的責任，使 Service 與 Pod 之間的溝通都是以無加密方式傳輸，得以正常傳輸資料。
- 支持虛擬網域設定（Offer Name Based Virtual Hosting）：Ingress 提供我們在同個 IP 下設定自己的虛擬網域，也就是我們前面提到的 hostname。
- 針對路徑的請求分配（Path-Based Routing）：除了根據 hostname 和 pathname 將請求導向特定的 Services，Ingress 還可以根據特定的請求路徑來轉發請求。這種功能可以用於在單一服務中處理多種不同的路徑請求。
- 整合其他 Kubernetes 特性：Ingress 可以和 Kubernetes 中的其他元件（例如 Secrets 或 ConfigMaps）進行整合。例如，你可以將 SSL 憑證和私鑰保存在 Kubernetes 的 Secret 中，然後在 Ingress 設定中引用這些 Secret。
- 客製化錯誤頁面：Ingress 支援客製化錯誤頁面，這樣你就可以對錯誤回應提供自定義的處理方式。
- 集成驗證系統：Ingress 可以結合第三方驗證系統，例如 JWT 或 OAuth，為應用提供更安全的訪問管控。

額外需要注意的是，不是所有的 Ingress 控制器都支援上述的所有特性，實際應用環境中使用的 Kubernetes 平台，在實現方式可能會有些許差異，我們需要根據具體的 Ingress 控制器的檔案來確定其支援的特性。

▷ 10.3　Ingress 安裝

在本地的 docker-desktop 上我們只需要運行下列設定檔，Kubernetes 就會幫我們建立一個 ingress-nginx 的 namespace，並且運行起相關服務：

```
1.  kubectl apply -f https://raw.githubusercontent.com/kubernetes/
    ingress-nginx/controller-v1.2.1/deploy/static/provider/cloud/
    deploy.yaml
2.  -------------------------------
3.  namespace/ingress-nginx configured
4.  serviceaccount/ingress-nginx configured
5.  serviceaccount/ingress-nginx-admission configured
6.  role.rbac.authorization.k8s.io/ingress-nginx configured
7.  role.rbac.authorization.k8s.io/ingress-nginx-admission configured
8.  clusterrole.rbac.authorization.k8s.io/ingress-nginx configured
9.  clusterrole.rbac.authorization.k8s.io/ingress-nginx-admission
    configured
10. rolebinding.rbac.authorization.k8s.io/ingress-nginx configured
11. rolebinding.rbac.authorization.k8s.io/ingress-nginx-admission
    configured
12. clusterrolebinding.rbac.authorization.k8s.io/ingress-nginx
    configured
13. clusterrolebinding.rbac.authorization.k8s.io/ingress-nginx-
    admission configured
14. configmap/ingress-nginx-controller configured
15. service/ingress-nginx-controller created
```

```
16. service/ingress-nginx-controller-admission created
17. deployment.apps/ingress-nginx-controller created
18. job.batch/ingress-nginx-admission-create created
19. job.batch/ingress-nginx-admission-patch created
20. ingressclass.networking.k8s.io/nginx configured
21. validatingwebhookconfiguration.admissionregistration.k8s.io/
    ingress-nginx-admission configured
```

檢查是否成功運作：

```
1.  kubectl get all -n ingress-nginx
2.  -----------------------------
3.  NAME                                         READY  STATUS     RESTARTS AGE
4.  pod/ingress-nginx-admission-create-rf8cl     0/1 Completed  0      7m4s
5.  pod/ingress-nginx-admission-patch-mzmc8      0/1 Completed  0      7m4s
6.  pod/ingress-nginx-controller-778667bc4b-twt6n 1/1  Running 0      7m4s
7.
8.  NAME                            TYPE          CLUSTER-IP
    EXTERNAL-IP    PORT(S)                  AGE
9.  service/ingress-nginx-controller           LoadBalancer
    10.105.184.158   localhost     80：30205/TCP,443：31820/TCP   7m5s
10. service/ingress-nginx-controller-admission   ClusterIP
    10.106.69.252    <none>        443/TCP                  7m4s
11.
12. NAME                            READY UP-TO-DATE AVAILABLE AGE
13. deployment.apps/ingress-nginx-controller 1/1    1      1       7m4s
14.
15. NAME                            DESIRED   CURRENT   READY   AGE
16. replicaset.apps/ingress-nginx-controller-778667bc4b 1   1  1  7m4s
17.
18. NAME                            COMPLETIONS   DURATION   AGE
19. job.batch/ingress-nginx-admission-create    1/1      5s       7m4s
20. job.batch/ingress-nginx-admission-patch     1/1      5s       7m4s
```

⤜ 10.4 實戰演練

關於 Ingress 的實際應用，官方有提供幾種方式讓我們用 URL 控制並連接到我
們指定的服務。

1. 單一 Service

現有的 Kubernetes 允許我們直接暴露單一個 Service 。現在我們依然可以透過
ingress 的 defaultBackend 辦到這件事，代表規範條件以外的流量通通都會遵
守 defaultBackend 規則分配到對應的服務。

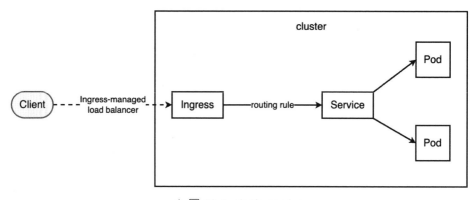

▲ 圖 10-1 single service

準備相關設定檔：

範例檔 **single-service / deployment.yaml**

```
1.  # single-service/deployment.yaml
2.  apiVersion: apps/v1
3.  kind: Deployment
4.  metadata:
5.    name: foo-deployment
6.    labels:
```

```
7.        type: demo
8.   spec:
9.     replicas: 2
10.    selector:
11.      matchLabels:
12.        type: demo
13.    template:
14.      metadata:
15.        labels:
16.          type: demo
17.      spec:
18.        containers:
19.        - name: foo
20.           image: mikehsu0618/foo
21.           ports:
22.           - containerPort: 8080
```

範例檔 **single-service / service.yaml**

```
1.   # single-service/service.yaml
2.   apiVersion: v1
3.   kind: Service
4.   metadata:
5.     name: my-service
6.   spec:
7.     selector:
8.       type: demo
9.     type: NodePort  # 這裡我們將能直接暴露連接埠的 'Loadbalancer' 改成
     'NodePort'
10.    ports:
11.    - protocol: TCP
12.        port: 8000
```

```
13.        targetPort: 8080
14.        nodePort: 30390
```

範例檔 **single-service / ingress.yaml**

```
1.  # single-service/ingress.yaml
2.  apiVersion: networking.k8s.io/v1
3.  kind: Ingress
4.  metadata:
5.    name: my-ingress
6.  spec:
7.    ingressClassName: nginx
8.    defaultBackend:
9.      service:
10.       name: my-service
11.       port:
12.         number: 8000
```

在上面的 ingress.yaml 我們設定了 defaultBackend 讓所有流量都預設導到 my-service 中的 8000 port，形成了一條從 loadbalancer → services → pods 的路徑。

執行以上設定檔：

```
1.  kubectl apply -f deployment.yaml,service.yaml,ingress.yaml
2.  --------------------------
3.  deployment.apps/foo-deployment unchanged
4.  service/my-service unchanged
5.  ingress.networking.k8s.io/my-ingress unchanged
```

查看服務狀況：

```
1.  kubectl get all
2.  --------------------------
```

```
3.  NAME                                READY    STATUS    RESTARTS    AGE
4.  pod/foo-deployment-6bbf665b47-6769n    1/1      Running   0           41m
5.  pod/foo-deployment-6bbf665b47-96khw    1/1      Running   0           41m
6.
7.  NAME               TYPE        CLUSTER-IP     EXTERNAL-IP    PORT(S)    AGE
8.  service/kubernetes    ClusterIP   10.96.0.1      <none>     443/TCP    27d
9.  service/my-service    NodePort    10.108.203.7   <none>     8000 : 30390/
    TCP    41m
10.
11. NAME                              READY    UP-TO-DATE    AVAILABLE   AGE
12. deployment.apps/foo-deployment    2/2      2             2           41m
13.
14. NAME                                    DESIRED   CURRENT   READY   AGE
15. replicaset.apps/foo-deployment-6bbf665b47    2       2        2      41m
```

查看 Ingress：

```
1.  kubectl get ingress
2.  ---------------------------
3.  NAME          CLASS    HOSTS    ADDRESS       PORTS    AGE
4.  my-ingress    nginx    *        localhost     80       42m
```

成功啟動了一個 localhost:80 的負載均衡器。

實際測試：

```
1.  curl localhost
2.  ---------------------------
3.  {"data": "Hello foo"}
```

2. Simple Fanout and Visual hosting

一個 fanout 可以根據請求的 URL 將來自同一個 IP 位址的流量轉到多個 Service。
並且實現以下設定：

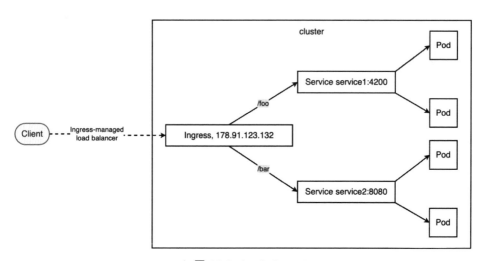

▲ 圖 10-2 simple fanout

準備相關設定檔：

範例檔 **simple-fanout / deployment.yaml**

```
1.   # simple-fanout/deployment.yaml
2.   apiVersion: apps/v1
3.   kind: Deployment
4.   metadata:
5.     name: foo-deployment
6.     labels:
7.       type: foo-demo
8.   spec:
9.     replicas: 2
10.    selector:
11.      matchLabels:
```

```
12.        type: foo-demo
13.    template:
14.      metadata:
15.        labels:
16.          type: foo-demo
17.      spec:
18.        containers:
19.          - name: foo
20.            image: mikehsu0618/foo
21.            ports:
22.              - containerPort: 8080
23. ---
24. apiVersion: apps/v1
25. kind: Deployment
26. metadata:
27.    name: bar-deployment
28.    labels:
29.      type: bar-demo
30. spec:
31.    replicas: 1
32.    selector:
33.      matchLabels:
34.        type: bar-demo
35.    template:
36.      metadata:
37.        labels:
38.          type: bar-demo
39.      spec:
40.        containers:
41.          - name: bar
42.            image: mikehsu0618/bar
43.            ports:
44.              - containerPort: 8080
```

範例檔 **simple-fanout / service.yaml**

```
1.  # simple-fanout/service.yaml
2.  apiVersion: v1
3.  kind: Service
4.  metadata:
5.    name: foo-service
6.  spec:
7.    type: NodePort
8.    selector:
9.      type: foo-demo
10.   ports:
11.     - protocol: TCP
12.       port: 8000
13.       targetPort: 8080
14. ---
15. apiVersion: v1
16. kind: Service
17. metadata:
18.   name: bar-service
19. spec:
20.   type: NodePort
21.   selector:
22.     type: bar-demo
23.   ports:
24.     - protocol: TCP
25.       port: 8000
26.       targetPort: 8080
```

範例檔 **simple-fanout / ingress.yaml**

```
1.  # simple-fanout/ingress.yaml
2.  apiVersion: networking.k8s.io/v1
```

```
3.   kind: Ingress
4.   metadata:
5.     name: my-ingress
6.   spec:
7.     ingressClassName: nginx
8.     rules:
9.       - host: foo.com
10.        http:
11.          paths:
12.            - path: /
13.              pathType: Prefix
14.              backend:
15.                service:
16.                  name: foo-service
17.                  port:
18.                    number: 8000
19.      - host: bar.com
20.        http:
21.          paths:
22.            - path: /
23.              pathType: Prefix
24.              backend:
25.                service:
26.                  name: bar-service
27.                  port:
28.                    number: 8000
```

可以在上面的設定檔看到我們用 ingress 產生出的 virsual hosting，並且使不同的網域對應到不同的 Service，實現預期的 fanout 效果。

讓我們運行前面的設定檔：

```
1.   kubectl apply -f deployment.yaml,service.yaml,ingress.yaml
```

```
2.  --------------------------
3.  deployment.apps/foo-deployment configured
4.  deployment.apps/bar-deployment configured
5.  service/foo-service configured
6.  service/bar-service configured
7.  ingress.networking.k8s.io/my-ingress configured
```

查看 ingress 詳細資訊：

```
1.  kubectl describe ingress my-ingress
2.  --------------------------
3.  Name:              my-ingress
4.  Labels:            <none>
5.  Namespace:         default
6.  Address:           localhost
7.  Ingress Class:     nginx
8.  Default backend:   <default>
9.  Rules:
10.   Host        Path  Backends
11.   ----        ----  --------
12.   foo.com
13.             /   foo-service: 8000 (10.1.1.169: 8080,10.1.1.170: 8080)
14.   bar.com
15.             /   bar-service: 8000 (10.1.1.168: 8080)
16. Annotations:  <none>
17. Events:
18.   Type     Reason  Age                From                     Message
19.   ----     ------  ----               ----                     -------
20.   Normal   Sync    7m35s (x4 over 15m)  nginx-ingress-controller
    Scheduled for sync
```

Ingress 成功地建立了 foo.com 和 bar.com 兩個虛擬網域，這兩個網域已經被設定為與相對應的 Service 相連接。

然而，這兩個虛擬網域都設定在我們的本地環境上。因此，我們需要設定讓這兩個網域的請求都被導向到本地 IP（即 127.0.0.1）。我們可以透過修改 /etc/hosts 檔案來實現這一設定，這樣才能保證在本地環境下順利完成請求的路由：

```
1.   sudo vim /etc/hosts
2.
3.   // 在檔案中插入以下需要映射的網域
4.   127.0.0.1 foo.com
5.   127.0.0.1 bar.com
6.
7.   // 在鍵盤中手動輸入下列字元來儲存！
8.   :wq!
```

接著來實際測試：

```
1.   curl http://foo.com
2.   {"data": "Hello foo"}
3.
4.   curl http://bar.com
5.   {"data": "Hello bar"}
```

大功告成！

筆者碎碎念

感謝願意看到這裡的讀者，到這裡可以說是已經初窺 Kubernetes 的門徑，熟悉 Docker 的人已經有能力可以在本地 run 起自己想要的服務，並且設定 URL 路徑實現負載均衡。說說我自己的收穫，因為接觸了 Kubernetes 讓我開始學習思考如何實現一套微服務系統，它的出現對我這個之前總是在寫單體式應用的小白來說，對分散式架構有個更清晰的輪廓並且深深著迷，還有太多東西可以學習了，就讓我們繼續堅持下去吧。

參考資料

- Kubernetes Documentation-ingress
  https://kubernetes.io/docs/concepts/services-networking/ingress/

- [Kubernetes] Resource Object 概觀
  https://godleon.github.io/blog/Kubernetes/k8s-CoreConcept-
  ResourceObject-Overview/

- Kubernetes 那些事 ─ Ingress 篇（一）
  https://medium.com/andy-blog/kubernetes-%E9%82%A3%E4%BA
  %9B%E4%BA%8B-ingress-%E7%AF%87-%E4%B8%80-92944d4bf97d

CHAPTER

11

Kubernetes —
Pod 的生命週期

恭喜堅持到這裡的各位，走完前面幾篇實戰演練已經可以大聲的説：媽我在用 Kubernetes 了，但凡事一定都不會如想像中的單純，接下來我們要正式進入進階基礎篇，除了帶各位掌握最基本的知識，還將深入探討背後的原理。

不難發現前面提到的 Deployment、Service 都是以 Pod 為中心去實現各種需求，由此可知 Pod 的重要性，通常 Pod 可以被視為一個服務的單位，包含著一個或一個以上的容器，Pod 的狀態和配置決定了服務是否能夠順利運行，所以我們最好在一開始就掌握 Pod 的生命週期中每個不同的階段，並且使用各種方法使其保持預期的健康，像是存活 / 就緒探針、重啟策略等。

比如説，存活探針（Liveness Probe）可以讓 Kubernetes 了解何時需要重啟 Pod 的某個容器，而就緒探針（Readiness Probe）則告訴 Kubernetes 何時可以開始將流量引導到這個 Pod。這些都是保證我們服務穩定運行的重要工具，我們將在接下來的內容中詳細介紹。

⫸ 11.1 Pod 的生命週期

Pod 的生命週期可以被理解為從建立到退出的完整流程，期間經歷了各種不同的狀態變化。圖 11-1 為我們呈現了一個 Pod 的完整生命週期，其中涵蓋了我們的主容器（Main Container）、初始化容器（Init Container）、生命週期鉤子（Post Start / Pre Stop Hook），以及健康檢查機制（Liveness / Readiness Probe）。

在接下來的部分，我們將會深入介紹這些關鍵元素對 Pod 生命週期的影響。然而在此之前，我們需要先對 Pod 的狀態有更深的理解。狀態定義作為最頂層的狀態顯示，可以為我們簡單地反映出 Pod 當前的實際狀態，同時在遇到問題需要排錯時，它也是我們首先分析的地方。

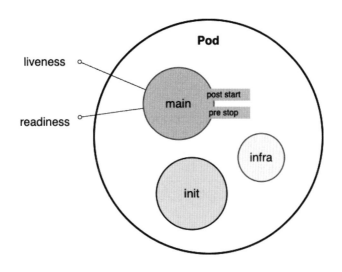

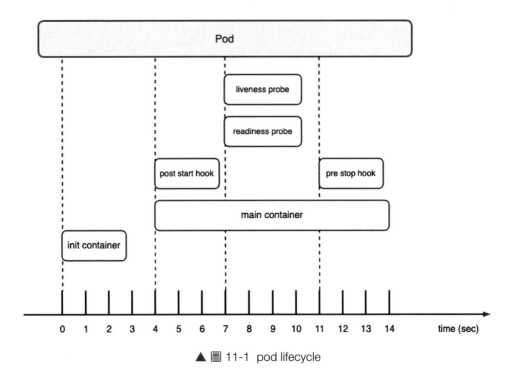

▲ 圖 11-1 pod lifecycle

11.2 Pod Phase（階段）

```
1.   kubectl get pods
2.   --------
3.   NAME                    READY     STATUS          RESTARTS          AGE
4.   admin-server            1/1       Running         0                 37h
5.   apps                    1/1       Running         0                 22d
```

這裡指的 Pod Phase 就是我們在查看 Pods 列表所用的 kubectl get pods 所帶出的 STATUS 欄位。

Pod Phase 所包含的狀態數量和定義是嚴格指定的，下面是 phase 可能的值：

- Pending：Pod 資訊已經提交給了叢集，但是還沒有被調度器調度到合適的節點或者 Pod 裡的鏡像正在下載。
- Running：該 Pod 已經綁定到了一個節點上，Pod 中所有的容器都已被建立。至少有一個容器正在運行，或者正處於啟動或重啟狀態。
- Succeeded：Pod 中的所有容器都被成功終止，並且不會再重啟。
- Failed：Pod 中的所有容器都已終止了，並且至少有一個容器是因為失敗終止。也就是說，容器以非「0」的狀態碼退出或者被系統終止。
- Unknown：因為某些原因無法取得 Pod 的狀態，通常是因為與 Pod 所在主機通訊失敗導致。

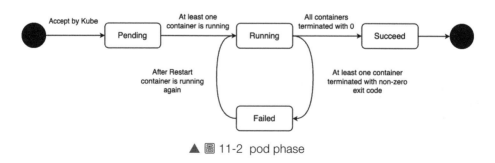

▲ 圖 11-2 pod phase

⯈ 11.3 重啟策略（Restart Policy）

我們可以透過設定 spec.template.spec.restartPolicy 來設置 Pod 中所有容器的重啟策略，其可能值為 Always，OnFailure 和 Never，預設值為 Always。容器的應用程式發生錯誤，或容器申請超出限制的資源都可能導致 Pod 終止，此時會根據 restartPolicy 來決定是否該重建 Pod。以下為三種可選的重啟策略：

- Always：Pod 終止就重啟，此為 default 設定。
- OnFailure：Pod 發生錯誤時才重啟。
- Never：從不重啟。

restartPolicy 僅只透過 kubelet 在同一節點上重新啟動容器。透過 kubelet 將不健康的容器，以指數增加延遲（10s，20s，40s…）重新啟動，上限為 5 分鐘，並在成功執行 10 分鐘後重置。不同類型的控制器可以控制 Pod 的重啟策略：

- Job：適用於一次性任務如批量計算，任務結束後 Pod 會被此類控制器清除。Job 的重啟策略只能是「OnFailure」或者「Never」。
- Replication Controller、ReplicaSet、Deployment：此類控制器希望 Pod 一直運行下去，它們的重啟策略只能是「Always」。
- DaemonSet：每個節點上只會啟動一個 Pod，很明顯此類控制器的重啟策略也應該是「Always」。

```
1.  apiVersion: apps/v1
2.  kind: Deployment
3.  metadata:
4.    name: my-app
5.    labels:
6.      app: my-app
7.  spec:
8.    serviceName: my-app
9.    replicas: 1
```

```
10.    selector:
11.      matchLabels:
12.        app: my-app
13.    template:
14.      metadata:
15.        labels:
16.          app: my-app
17.      spec:
18.        restartPolicy: Always
19.        containers:
20.        - name: my-app
21.          image: myregistry: 443/mydomain/my-app
22.          imagePullPolicy: Always
```

11.4 初始化容器（Init Container）

瞭解了 Pod 的狀態以及重啟策略後，接下來我們要看的是在 Pod 生命週期中最先啟動的 Init Container。顧名思義，Init Container 是在主容器運行之前會先行執行完畢的初始程式，可以是一個或多個。如果一個以上的話，這些容器將會按照定義的順序執行。我們知道，在一個 Pod 中，可以在所有容器中共享資料和 Network Namespace，所以我們可以利用初始化容器執行初始化動作，使一切資源就緒時再啟動主容器，這樣有益於我們將初始化的邏輯從主容器中解耦出來，變得更加靈活運用。那麼初始化容器還有哪些應用場景呢？

- 等待其他服務就緒：這種做法特別適用解決服務之間的依賴問題，比如說我們有個主服務依賴於另一個資料庫服務，但是在我們啟動這個主服務時，並不能保證被依賴的資料庫是否就緒，這時候我們可以簡單使用一個 Init Container 去監測資料庫是否就緒，確認就緒後就能直接退出，並且主程式將會在這時候接著啟動。

● 做初始化設定：比如叢集裡檢測所有已經存在的成員節點，為主容器準備好
  叢集的設定資訊，這樣主容器啟動後就能用這個設定資訊加入叢集。

現在來實現一個初始化容器去預先準備首頁內容：

```
1.   apiVersion: v1
2.   kind: Pod
3.   metadata:
4.     name: init-demo
5.   spec:
6.     volumes:
7.     - name: workdir
8.       emptyDir: {}
9.     initContainers:
10.    - name: install
11.      image: busybox
12.      command:
13.      - wget
14.      - "-O"
15.      - "/work-dir/index.html"
16.      - http://www.google.com
17.      volumeMounts:
18.      - name: workdir
19.        mountPath: "/work-dir"
20.    containers:
21.    - name: nginx
22.      image: nginx
23.      ports:
24.      - containerPort: 80
25.      volumeMounts:
26.      - name: workdir
27.        mountPath: /usr/share/nginx/html
```

可以簡單看出 initContainers 產生出 index.html 隨即退出，並且利用 volume 將資料夾目錄掛載到主容器中，實現初始化並共享資源的概念。

▸ 11.5 生命週期鉤子（Lifecycle Hook）

生命週期鉤子是在初始化容器執行完畢後由 kubelet 發起的，並且會跟著主程式一起啟動，並且在容器啟動時以及容器終止之前運行，而我們可以為 Pod 中的所有容器設定生命週期鉤子。

Kubernetes 為我們提供了兩種鉤子函數：

- PostStart：這個鉤子在容器建立後立即執行。但是並不能保證鉤子將在容器入口點（ENTRYPOINT）之前運行，因為它不會傳遞任何參數傳遞給作業程序。主要用於資源部署、環境準備等。不過需要注意的是如果鉤子花費太長時間以至於不能運行或者遇到問題，容器將不能達到 running 狀態。
- PreStop：是一個在容器結束運行前立刻被呼叫的鉤子。它是同步的，可以延遲終止過程，必須在容器被正式終止前完成執行。主要用於優雅的關閉應用、通知其他系統等。如果鉤子在執行期間掛起，Pod 階段將停留在 running 狀態，並且永不會達到 failed 狀態。

以下為設定生命週期鉤子的簡單範例：

```
1.  apiVersion: v1
2.  kind: Pod
3.  metadata:
4.    name: lifecycle-demo
5.  spec:
6.    containers:
7.    - name: lifecycle-demo-container
8.      image: nginx
```

```
9.       lifecycle:
10.        postStart:
11.          exec:
12.            command: ["/bin/sh", "-c", "echo Hello from the postStart
     handler > /usr/share/message"]
13.        preStop:
14.          exec:
15.            command: ["/bin/sh","-c","nginx -s quit; while killall -0
     nginx; do sleep 1; done"]
```

▶ 11.6 健康檢查（Health Check）

在主容器的整個運行生命週期中，健康檢查是能夠影響到 Pod 狀態的關鍵部分。在 Kubernetes 中我們可以透過各種探針來確認容器是否處於正常運作的狀態，像是存活探針（Liveness Probe）、就緒探針（Readiness Probe）和啟動探針（Startup Probe），如果出現異常將會透過自我檢測與修復來避免把流量導到不健康的 Pod。

Kubernetes 支援四種用在 Pod 探測的處理器：

- Exec：在容器內執行指令，再根據其回傳的狀態進行診斷，回傳 0 表示成功，其餘皆為失敗。
- TCPSocket：透過對容器上的 TCP 連接埠進行檢查，其連接埠有打開表示成功，否則為失敗。
- HTTPGet：透過對容器 IP 位址上的指定連接埠發起 http GET 請求進行診斷，如果回應狀態大於等於 200 且小於 400，則為成功，其餘皆為失敗。
- gRPC：從 v1.24 版本起，可以透過對容器 IP 位址上的指定連接埠發起 gRPC 請求進行診斷，這裡需要注意的是使用 gRPC 做為 action 時需要特別指定連接埠。

Kubelet 可以執行三種探測：

- livenessProbe：顯示容器是否正常運作，如果探測失敗 kubelet 會終止容器，容器會依照重啟策略進行下一個動作。如果容器不支援存活探測，則預設狀態為 Success。
- readinessProbe：顯示容器是否準備好提供服務，如果探測失敗，Endpoint Controller 會從匹配的所有 Service Endpoint 列表中刪除 Pod IP。如果容器不支援就緒探測，則預設狀態為 Success。
- startupProbe：顯示容器中的應用是否已經啟動，如果啟動 startupProbe 則其他探測都會被停用，直到 startupProbe 成功後其他探針才會開始接管，如果探測失敗 kubelet 會終止容器，容器會依照重啟策略進行下一個動作。如果容器不支援啟用探測，則預設狀態為 Success。

以下為探針設定的簡單範例：

```
1.  apiVersion: v1
2.  kind: Pod
3.  metadata:
4.    name: goproxy
5.    labels:
6.      app: goproxy
7.  spec:
8.    containers:
9.    - name: goproxy
10.     image: k8s.gcr.io/goproxy: 0.1
11.     ports:
12.     - containerPort: 8080
13.     readinessProbe:
14.       tcpSocket:
15.         port: 8080
16.       initialDelaySeconds: 5
```

```
17.        periodSeconds: 10
18.     livenessProbe:
19.       tcpSocket:
20.         port: 8080
21.       initialDelaySeconds: 15
22.       periodSeconds: 20
```

在 Kubernetes 叢集中，kubelet 是一個運行在每個節點（Node）上的主要行程。它的主要功能是負責管理該節點上的容器，包括啟動、停止，以及監控容器的狀態。換句話說，kubelet 的任務是確保每個節點上的容器都在運行並且處於正確的狀態。

livenessProbe、readinessProbe、以及 startupProbe，都是 kubelet 用來檢查和管理容器狀態的工具。例如，livenessProbe 用於檢查容器是否還在運行，如果探測檢查失敗（即容器沒有在運行），那麼 kubelet 就會終止該容器，並根據設定的重啟策略來決定是否重啟這個容器。這就是為什麼在討論這些探測時會提及 kubelet，因為 kubelet 是負責執行這些探測檢查，並根據檢查結果來管理容器的行程。

🛢筆者碎碎念

相信各位無論在學習前後端語言或是各種框架各種應用時，都能常常看到生命週期這個關鍵字眼，能夠理解並善用生命週期可以讓我們了解一個應用的一生中都經歷了些什麼，如此一來我們就能在正確的時間點，使其為我們執行更準確的動作，這會大幅提升應用的靈活性以及上限，所以我們當然需要了解被 Kubernetes 環繞建構而成的 Pod 的生命週期，是吧是吧！

參考資料

- Configure Liveness, Readiness and Startup Probes
  https://kubernetes.io/docs/tasks/configure-pod-container/configure-liveness-readiness-startup-probes/

- Pod 的生命週期
  https://jimmysong.io/kubernetes-handbook/concepts/pod-lifecycle.html

- POD 生命週期
  https://ithelp.ithome.com.tw/articles/10243067

- day 10 Pod(3)- 生命週期 , 容器探測
  https://ithelp.ithome.com.tw/articles/10236314

- API OVERVIEW
  https://kubernetes.io/docs/reference/generated/kubernetes-api/v1.25/#probe-v1-core

- restartPolicy: Unsupported value: "Never": supported values: "Always"
  https://stackoverflow.com/questions/55169075/restartpolicy-unsupported-value-never-supported-values-always

Kubernetes Kubectl 指令 與它的快樂夥伴

在 Kubernetes 的世界中，kubectl 工具無疑扮演著極為重要的角色。作為 kube-apiserver 的命令列介面，kubectl 使得我們能夠與 Kubernetes 叢集進行直接對話。不僅如此，kubectl 的功能遠遠超出了僅僅對 Kubernetes 資源進行基本的 CRUD 操作。

一方面，kubectl 支援各種強大的子指令，包括日誌查看（log）、節點和服務狀態監控（top）、實時追蹤資源變化（watch）等等，極大地方便了我們對 Kubernetes 叢集的維運管理工作。另一方面，kubectl 提供了一套完善的 YAML / JSON 操作介面，使得我們可以在一個統一的框架下對 Kubernetes 資源進行複雜的設定和調度。

此外，瞭解 kubectl 的使用也有助於我們理解 Kubernetes 的底層運作機制。從指令的選項參數，到 API 版本和資源類型，再到錯誤訊息和日誌輸出，都是 Kubernetes 運作原理的有力體現。因此，熟練掌握 kubectl 不僅可以提高我們對 Kubernetes 的操作效率，也對理解 Kubernetes 的設計哲學和實作細節大有益處。

因此，無論你是 Kubernetes 的初學者還是有經驗的開發者，都應該花時間去熟悉並掌握 kubectl。不僅可以幫助你更順利地使用和管理你的 Kubernetes 叢集，也可以進一步提升你對雲端原生技術的理解和掌握。

12.1 Kubectl 介紹

kubectl 是 Kubernetes 的命令列介面（Command Line Interface，CLI），讓我們可以透過命令列對 Kubernetes 叢集進行操作。

從部署應用程式、檢查資源狀態、到除錯問題和編輯資源，kubectl 提供了多種強大的功能，讓我們以簡單直觀的方式與 Kubernetes 進行互動。其功能包含但不限於：

- 查詢 Kubernetes 叢集的狀態：我們可以使用 kubectl get 指令來查詢叢集中各種資源（例如 Pods、Services、Deployments 等）的狀態。
- 修改 Kubernetes 資源：我們可以透過 kubectl edit 或 kubectl apply 來修改現有的 Kubernetes 資源。
- 部署和管理應用程式：kubectl run 或 kubectl create 可以用來部署新的應用程式，而 kubectl scale 或 kubectl autoscale 可以用來調整應用程式的運行規模。
- 疑難排解和除錯：當你需要檢查問題或了解 Kubernetes 資源的詳細訊息時，kubectl describe 或 kubectl logs 可以提供你需要的資訊。

這些功能讓 kubectl 成為 Kubernetes 的重要操作工具，幫助我們在進行 Kubernetes 任務時提供一個方便而直觀的操作方式。在後續的內容中，我們將深入了解 kubectl 的各種使用方法和技巧，以幫助你在使用 Kubernetes 時能更加得心應手。

⯈ 12.2 Kubectl 安裝設定

安裝和設定 kubectl 是使用 Kubernetes 的第一步，kubectl 可以在多種操作系統上進行安裝，包括 Linux、macOS 和 Windows。以下示範在 macOS 系統安裝的操作步驟。

在 macOS 上，可以使用 Homebrew 進行安裝：

```
1.   brew install kubectl
```

一旦 kubectl 安裝完成，需要設定與 Kubernetes 叢集的連接。

最常見的方式是透過 kubeconfig 檔案（預設位於你的 home 目錄下的 .kube/config）來設定 kubectl。如果你使用的是 Google Kubernetes Engine（GKE）、Amazon EKS 或 Azure AKS 這樣的雲端服務，它們的命令列工具會為你生成適當的 kubeconfig 檔案。如果你想手動設定 kubeconfig，可以參考 Kubernetes 官方文件中的設定指南。

安裝並設定好 kubectl 後，就可以開始控制和管理 Kubernetes 叢集了。使用 kubectl get nodes 指令，你應該能看到 Kubernetes 叢集中的節點列表，這代表你已經成功設定並連接到叢集了。

至此，我們已經成功安裝並設定好 kubectl，接下來可以開始學習和探索更多 Kubernetes 的功能了。

12.3 Kubectl 語法

如果想要熟悉 kubectl 語法，應該深入了解每個指令的功能和語法。隨著經驗的累積，你將能夠根據已有知識推斷新的指令操作，從而真正將這些技能內化。

kubectl 的語法如下：

```
1.  kubectl [command] [type] [name] [flags]
```

使用以下語法從終端窗口執行 kubectl 指令，其中 command、type、name 和 flags 分別是：

- command：指定要對一個或多個資源執行的操作，例如 create、get、describe、delete。
- type：指定資源類型。資源類型不區分大小寫，可以指定單數、複數或縮寫形式。例如，以下指令輸出相同的結果：
  - 取得 Pod：

```
1.  kubectl get pod pod1
2.  kubectl get pods pod1
3.  kubectl get po pod1
```

- 取得 Service：

```
1.  kubectl get service service1
2.  kubectl get services service1
3.  kubectl get svc service1
```

- 取得 Deployment：

```
1.  kubectl get deployment deployment1
2.  kubectl get deployments deployment1
3.  kubectl get deploy deployment1
```

- name：指定資源的名稱。名稱區分大小寫。如果省略名稱，則顯示所有資源的詳細資訊。例如：kubectl get pods。

  1. 在對多個資源執行操作時，你可以按類型和名稱指定每個資源，或指定一個或多個檔案。
  2. 要按類型和名稱指定資源。
  3. 要對所有類型相同的資源進行分組，請執行這個操作：type name1 name2。

     例子：kubectl get pod example-pod1 example-pod2
  4. 分別指定多個資源類型：type1/name1 type1/name2 type3/name3。

     例子：kubectl get pod/example-pod1 replicationcontroller/example-rc1
  5. 用一個或多個檔案指定資源。

     例子：-f file1 -f file2
  6. 使用 YAML 而不是 JSON，因為 YAML 對使用者更友好，特別是對於設定檔案。例子：kubectl get -f ./pod.yaml

- flags：指定可選的參數。例如，可以使用 -s 或 --server 參數指定 Kubernetes API 伺服器的位址和連接埠。

  注意：從命令列指定的參數會覆蓋預設值和任何相應的環境變數。

▶ 12.4 Kubectl 常用指令

現在我們將介紹一些最基本和最常用的 kubectl 指令，讓你更有效地管理和操作你的 Kubernetes 叢集。這些指令都是我們身為 Kubernetes 使用者必須知道的。

- apply：以檔案或標準輸入為準應用或更新資源。

```
1.  # 使用 example-service.yaml 建立服務
2.  kubectl apply -f example-service.yaml
3.
4.  # 使用 <directory> 路徑下的任意 .yaml、.yml 或 .json 檔案建立物件
5.  kubectl apply -f <directory>
```

- describe：顯示一個或多個資源的詳細狀態，預設情況下包括未初始化的資源。

```
1.  # 顯示名為 <node-name> 的 Node 的詳細資訊。
2.  kubectl describe nodes <node-name>
3.
4.  # 顯示名為 <pod-name> 的 Pod 的詳細資訊。
5.  kubectl describe pods/<pod-name>
6.
7.  # 顯示由名為 <rc-name> 的副本控制器管理的所有 Pod 詳細資訊。
8.  # 記住：副本控制器建立的任何 Pod 都以副本控制器的名稱為前綴。
9.  kubectl describe pods <rc-name>
10.
11. # 描述所有的 Pod
12. kubectl describe pods
```

- get：用於獲取叢集的一個或一些 resource 訊息。該指令可以列出叢集所有資源的詳細資訊，resource 包括叢集節點、運行的 Pod、Deployment、Service 等。

> 叢集中可以建立多個 Namespace，未指定 Namespace 的情況下，所有操作都是
> 針對 --namespace=default，而 --all-namespaces／-A 則可以把範圍擴大到全部
> 的 Namespace。

例如：

```
1.  # 獲取 default namespace 所有 pod 的詳細資訊：
2.  kubectl get po -o wide
3.
4.  # 獲取所有 namespace 下的運行的所有 pod：
5.  kubectl get po --all-namespaces
6.
7.  # 獲取 default namespace 下的運行的所有 pod 的標籤：
8.  kubectl get po --show-labels
9.
10. # 獲取該叢集的所有命名空間：
11. kubectl get namespace
```

- create：根據檔案或者輸入來建立資源。

```
1.  kubectl create -f demo-deployment.yaml
2.  kubectl create -f demo-service.yaml
```

- delete：刪除資源。

```
1.  kubectl delete -f demo-deployment.yaml
2.  kubectl delete -f demo-service.yaml
3.  kubectl delete <rc> <rc-name>
```

- run：在叢集中建立並運行一個或多個容器鏡像。

```
1.  # 語法
2.  kubectl run NAME --image=image [--env="key=value"] \
3.      [--port=port] [--replicas=replicas] [--dry-run=bool] \
4.      [--overrides=inline-json] [--command] -- [COMMAND] [args...]
```

```
5.  # 運行一個名稱為 nginx，副本數為 3，標籤為 app=example，鏡像為 nginx：
    1.10，連接埠為 80 的容器實例
6.  kubectl run nginx --replicas=3 --labels="app=example"
    --image=nginx:1.10 --port=80
```

● expose：建立一個 service 服務，並且暴露連接埠讓外部可以訪問。

```
1.  # 建立一個 nginx 服務並暴露 88 連接埠讓外部訪問
2.  kubectl expose deployment nginx --port=88 --type=NodePort --target-
    port=80 --name=nginx-service
```

● set：設定應用的一些特定資源，也可以修改應用已有的資源。

```
1.  # 使用 kubectl set --help 查看其可用的子指令，env，image，resources，
    selector，serviceaccount，subject。
2.
3.  # 語法
4.  kubectl resources (-f FILENAME | TYPE NAME) ([--limits=LIMITS &
    --requests=REQUESTS]
```

● exec：對 Pod 中的容器執行命令。

```
1.  # 從 Pod <pod-name> 中獲取運行 'date' 的輸出。預設情況下，輸出來自第一個
    容器。
2.  kubectl exec <pod-name> -- date
3.
4.  # 運行輸出'date'獲取在 Pod <pod-name> 中容器 <container-name> 的輸出。
5.  kubectl exec <pod-name> -c <container-name> -- date
6.
7.  # 獲取一個互動 TTY 並在 Pod <pod-name> 中運行 /bin/bash。預設情況下，輸
    出自第一個容器。
8.  kubectl exec -ti <pod-name> -- /bin/bash
```

● logs：印出 Pod 中容器的日誌。

```
1.  # 回傳 Pod <pod-name> 的日誌快照。
```

```
2.  kubectl logs <pod-name>
3.
4.  # 從 Pod <pod-name> 開始流式傳輸日誌。這種類似於 'tail -f' Linux 指令。
5.  kubectl logs -f <pod-name>
```

接下來我們再介紹一些進階使用方法：

- kubectl autocompletion：你可以啟用 kubectl 的自動補全功能，以更有效地在命令列工具中使用 kubectl。如果你正在使用 bash，你可以嘗試以下指令：

```
1.  source <(kubectl completion bash)
```

將此指令添加到你的 .bashrc 或 .bash_profile 檔案中，以便每次打開新的終端機時都會啟用自動補全。

- kubectl config 使用：你可以使用 kubectl config 指令來管理你的 kubectl 設定，這在操作多個 Kubernetes 叢集時特別有用。例如，你可以使用以下指令切換到一個不同的 Kubernetes 叢集：

```
1.  kubectl config use-context my-cluster-name
```

- kubectl patch：你可以使用 kubectl patch 指令來即時修改現有的 Kubernetes 物件，而不需要完全替換它們。例如，你可以更改一個現有 Deployment 的 Image：

```
1.  kubectl patch deployment my-deployment -p '{"spec":{"template":
    {"spec":{"containers":[{"name":"my-container","image":"my-new-
    image"}]}}}}'
```

- kubectl debug：你可以使用 kubectl debug 指令來啟動除錯 Pod。這將建立一個新的容器，該容器與你要除錯的 Pod 具有相同的設定，讓你能夠更輕鬆地進行除錯：

```
1.  kubectl debug my-pod
```

- kubectl logs --since：你可以使用 kubectl logs --since 來獲取 Pod 在特定時間段的日誌。例如，獲取過去一小時內的日誌：

```
1.   kubectl logs my-pod --since=1h
```

- kubectl port-forward：你可以使用 kubectl port-forward 將你的本地連接埠轉發到運行在 Kubernetes 中的 Pod。這樣可以讓你在本地輕鬆訪問該 Pod 提供的服務：

```
1.   kubectl port-forward my-pod 8080:8080
```

- 使用 jsonpath 擷取特定資訊：除了 kubectl get 指令的 -o json 選項能夠讓你看到 Kubernetes 物件的完整資訊外，你還可以使用 jsonpath 運算式來擷取你關心的特定資訊。例如，你可以使用以下指令來獲得一個 Pod 的 IP 位址：

```
1.   kubectl get pod my-pod -o=jsonpath='{.status.podIP}'
```

- 使用 label 選擇器：在 Kubernetes 中，labels 是一種很重要的方式來管理和選擇物件。你可以使用 -l 或 --selector 選項來根據 label 選擇物件。例如，這個指令可以選擇所有標有 app=frontend 的 Pods：

```
1.   kubectl get pods -l app=frontend
```

- 使用 kubectl apply 的 --prune 選項：如果你有一組 Kubernetes 設定檔案，並且希望 Kubernetes 叢集的狀態能夠與這些設定檔案同步，你可以使用 kubectl apply 的 --prune 選項。這個選項將會刪除叢集中所有未在設定檔案中定義的物件：

```
1.   kubectl apply -f my-dir/ --prune -l app=frontend
```

- 使用 kubectl 的 --dry-run 選項：如果想要檢查一個操作但是又不想真的執行它，你可以使用 kubectl 的 --dry-run 選項。這將僅回傳該操作可能的結果，但並不會真的執行該操作：

```
1.   kubectl create deployment my-deployment --image=my-image --dry-
     run=client
```

▶ 12.5 善加利用 Kubectl Help

對於任何新的命令列工具,熟悉其內建的幫助功能是一個重要且實用的步驟。
在 kubectl 中,你可以透過 kubectl help 指令來獲取該工具提供的各種指令及
其使用方式的資訊。

當你在命令列輸入 kubectl help,將會印出一個基本的指令列表,包括了各種
操控 Kubernetes 資源、管理叢集狀態,以及診斷問題的命令。每個指令都有對
應的簡短說明,可供你快速理解該指令的功能。

此外,如果你想要進一步了解某個特定的 kubectl 指令,只需要在 kubectl help
後面加上你有興趣的指令名稱,例如 kubectl help get。這會印出該指令的詳細
說明、可用選項,以及使用範例。

透過 kubectl help,你可以掌握 kubectl 的大部分功能,並在遇到困難時尋找解
答。這項功能是 kubectl 的重要學習資源,我們鼓勵你充分利用這項工具,以
協助你探索 Kubernetes 的各種可能性。因此,當你在使用 kubectl 時,不要忘
記 kubectl help 這個良師益友。

```
1.   kubectl help
2.   -------------------
3.   kubectl controls the Kubernetes cluster manager.
4.    Find more information at: https://kubernetes.io/docs/reference/
     kubectl/overview/
5.
6.   Basic Commands (Beginner):
7.    create          Create a resource from a file or from stdin
```

```
8.    expose        Take a replication controller, service,
   deployment or pod and expose it as a new Kubernetes service
9.    run           Run a particular image on the cluster
10.   set           Set specific features on objects
11.
12. Basic Commands (Intermediate):
13.   explain       Get documentation for a resource
14.   get           Display one or many resources
15.   edit          Edit a resource on the server
16.   delete        Delete resources by file names, stdin, resources
   and names, or by resources and label selector
17.
18. Deploy Commands:
19.   rollout       Manage the rollout of a resource
20.   scale         Set a new size for a deployment, replica set, or
   replication controller
21.   autoscale     Auto-scale a deployment, replica set, stateful
   set, or replication controller
22.
23. Cluster Management Commands:
24.   certificate   Modify certificate resources.
25.   cluster-info  Display cluster information
26.   top           Display resource (CPU/memory) usage
27.   cordon        Mark node as unschedulable
28.   uncordon      Mark node as schedulable
29.   drain         Drain node in preparation for maintenance
30.   taint         Update the taints on one or more nodes
31.
32. Troubleshooting and Debugging Commands:
33.   describe      Show details of a specific resource or group of
   resources
34.   logs          Print the logs for a container in a pod
```

```
35.    attach         Attach to a running container
36.    exec           Execute a command in a container
37.    port-forward   Forward one or more local ports to a pod
38.    proxy          Run a proxy to the Kubernetes API server
39.    cp             Copy files and directories to and from containers
40.    auth           Inspect authorization
41.    debug          Create debugging sessions for troubleshooting
       workloads and nodes
42.
43. Advanced Commands:
44.    diff           Diff the live version against a would-be applied
       version
45.    apply          Apply a configuration to a resource by file name
       or stdin
46.    patch          Update fields of a resource
47.    replace        Replace a resource by file name or stdin
48.    wait           Experimental: Wait for a specific condition on
       one or many resources
49.    kustomize      Build a kustomization target from a directory
       or URL.
50.
51. Settings Commands:
52.    label          Update the labels on a resource
53.    annotate       Update the annotations on a resource
54.    completion     Output shell completion code for the specified
       shell (bash, zsh or fish)
55.
56. Other Commands:
57.    alpha          Commands for features in alpha
58.    api-resources  Print the supported API resources on the server
59.    api-versions   Print the supported API versions on the server,
       in the form of "group/version"
```

```
60.    config            Modify kubeconfig files
61.    plugin            Provides utilities for interacting with plugins
62.    version           Print the client and server version information
63.
64.  Usage:
65.    kubectl [flags] [options]
66.
67.  Use "kubectl <command> --help" for more information about a given
     command.
68.  Use "kubectl options" for a list of global command-line options
     (applies to all commands).
```

筆者碎碎念

到目前為止我們已經了解了基本的指令以及設定，之後的方向我們將逐漸揭開 Kubernetes 的神秘面紗，使用更實際以及更深入的例子來學習各種設定，以及了解該設定的存在的原因，kubectl 都給予我們全面而深入的掌握能力。

參考資料

- Kubernetes 教學系列 - kubectl 常見指令説明
  https://blog.kennycoder.io/2020/12/18/Kubernetes%E6%95%99%E5%AD%B8%E7%B3%BB%E5%88%97-kubectl%E5%B8%B8%E8%A6%8B%E6%8C%87%E4%BB%A4%E8%AA%AA%E6%98%8E/

- kubectl Cheat Sheet
  https://kubernetes.io/docs/reference/kubectl/cheatsheet/

- Command line tool (kubectl)
  https://kubernetes.io/docs/reference/kubectl/

Part 5
實戰部署篇

這些花式部署
你學會了嗎？

在現今的科技世界中，軟體的快速迭代已經成為
了一種常態。每個產品、每個服務都需要在短時
間內不斷的進行更新與優化，以應對市場變動和
用戶需求的挑戰。在這種情況下，一個有效且靈
活的部署策略成為了必不可少的工具。

Kubernetes Deployment Strategies — 常見的部署策略

如今在我們的實際工作環境中，產品的生命週期越來越短且迭代速度日益加快，身為 Server 守護者的我們可能會面臨到一兩週就要迭代一次，甚至是一天迭代數次的情況。我們需要知道大部分的技術創新價值始終來自於人們對它的了解以及使用程度，因此有效的部署策略也是技術創新的一個重要因素，不論是推出時機、授權或是行銷，一個適當的部署策略可以使我們一探市場接受程度、降低服務停運時間以及在機會成本中做出取捨。你可以觀察到一個好的部署策略應該要具備盡可能地讓服務不中斷、可以戰略性地試試市場水溫，以及擁有回溯到歷史版本的能力，接下來就來介紹幾種常見的部署策略以及它們的優劣。

▶ 13.1 重建部署（Recreate）

重建部署可以說是一種成本較高的部署方式。簡單來說，這種策略會先將舊版本的應用完全下線，然後再開始部署新版本的應用，這意味著在新舊版本切換期間，你的服務將會有一段時間無法提供服務，這段時間的長短取決於應用下線和啟動的耗時。

優點：

- 設定相對簡單明瞭。
- 在部署過程中，只會有一個版本的應用在運行。
- 在部署過程中，不會對主機產生額外的負擔。

缺點：

- 這種部署策略會對使用者造成顯著的影響，因為它會導致明確的服務中斷。此中斷的持續時間依賴於停止舊服務和啟動新服務的時間。

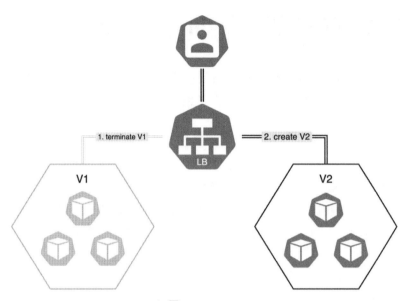

▲ 圖 13-1 recreate

假設我們有一個正在運行的 Deployment 名為「my-app」，現在我們需要將它更新到新版本。在進行重建部署時，我們首先需要將舊的「my-app」完全下線：

```
1.  kubectl scale deployment my-app --replicas=0
```

等到舊的「my-app」完全停止後，我們再將新版本的應用部署上去：

```
1.  kubectl set image deployment my-app my-app=new-version-image
2.  kubectl scale deployment my-app --replicas=3
```

其中，new-version-image 是新版本應用的 Docker Image。--replicas=3 是將新版本的應用部署至 3 個副本。

需要注意的是，這種部署方式會導致應用在舊版本停止到新版本啟動這段時間內無法提供服務，因此一般不推薦在生產環境中使用。而對於不可以有服務中斷的情況，我們通常會選用滾動更新（Rolling Update）或藍綠部署（Blue-Green Deployment）等更為平滑的部署策略。

13.2 滾動部署（Rolling Update）

滾動（Ramped，又稱為 Rolling Update）部署策略就是指容器會如同水漸漸地往傾斜的地方聚集一樣的更新版本，能緩慢平和地釋出新版本。

如同水流的流速快慢一樣，滾動部署也能透過調整以下引數來調整部署穩定性以及速率：

- 最大執行數：可以同時釋出的服務數目。
- 最大峰值：升級過程中最多可以比原先設定所多出的服務數量。
- 最大不可用數：最多可以有幾個服務處在無法服務的狀態。

優點：

- 相較於藍綠部署更加節省資源。
- 便於設定，服務不中斷。

缺點：

- 釋出與回滾耗時。想想如果我們有 100 個服務，每次需要花五分鐘更新其中 10 個，當更新到第 80 個時發現錯誤需要緊急回滾的情況？
- 部署期間新舊兩版服務都會同時在線上運作，無法控制流量且噴錯時，除錯的困難度較高。

在 Kubernetes 中，滾動更新被視為預設的部署策略。這意味著當你透過 Kubernetes 更新你的應用程式時，它會逐步替換舊的 Pod 實例，並以新的實例取代它們，這種滾動更新的方式確保了服務的高可用性和無縫更新。

這種策略的實現是透過 Kubernetes 的 Deployment 物件，它在後台管理和更新應用程式的 Pod。當你更改了 Deployment 的設定（例如更新了應用程式的 Image 版本）時，Deployment 會開始滾動更新。你可以透過設定 Deployment 的 spec 屬性（例如 .spec.strategy.type 和 .spec.strategy.rollingUpdate）來調整滾動更新的行為。

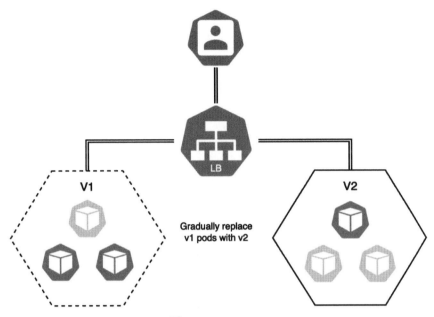

▲ 圖 13-2 rolling update

13.3 藍綠部署（Blue / Green）

相較於滾動更新，藍綠部署則是會先將新版本的服務完整的開啟，並且在新版本滿足上線條件的測試後，才將流量在負載均衡層從舊版本切換到新版本。

優點：

- 實時釋出、回滾。
- 避免新舊版本衝突，整個過程同時只會有一個版本存在。
- 服務不中斷。

缺點：

● 部署完成前，會因為需要雙倍的資源而增加額外的開銷和成本。有時新版本
  通過不了測試時，舊版本將持續運行到新版本通過為止。
● 當切換到新版本的瞬間，如果有未處理完成的業務將會是比較麻煩的問題。

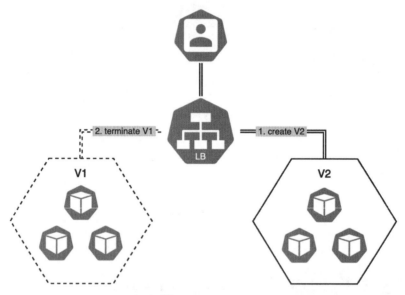

▲ 圖 13-3 blue green

藍綠部署策略是一種強大的部署策略，它允許開發團隊在完全分離的環境中同
時運行新舊版本，直到新版本經過充分的測試並準備好全面接管流量。這種策
略的優勢在於能夠實現快速且無縫的切換，如果新版本出現問題，能立即回滾
到上一版本。然而，這種策略也並非沒有挑戰。雙倍的資源需求可能對某些組
織來說是一個重大的障礙，而在切換時也需要注意正在處理的業務。

藍綠部署的最大優點是它將減少部署新版本可能帶來的風險。透過這種策略，
開發團隊可以確保新版本在完全上線之前，經過了充分的測試和驗證。因此，
對於尋求高度可用性和無縫升級體驗的組織來說，藍綠部署策略是一個值得考
慮的選擇。

⯈ 13.4 金絲雀部署（Canary）

金絲雀部署，與藍綠部署不同的是，它不是非黑即白的部署方式，所以又稱為灰度部署。

灰度部署是指在黑與白之間，能夠平滑過渡的一種部署方式。我們能夠緩慢的將新版本先推廣到一小部分的使用者，驗證沒有問題後才完成部署，以降低生產環境引入新功能帶來的風險。

例如將 90% 的請求導向舊版本，10% 的請求轉向新版本。這種部署大多用於缺少可靠測試或者對新版本穩定性缺乏信心的情況下。

NOTE

金絲雀部署的命名來自於 17 世紀的礦井工人發現金絲雀對瓦斯這種氣體非常敏感，哪怕是只有極其微量的瓦斯，金絲雀也會停止歌唱，率先比人類出現不良反應，所以工人每次下井時都會帶上一隻金絲雀作為危險狀況下的救命符。

優點：

- 方便除錯以及監控。
- 只向一小部分使用者釋出。
- 快速回滾、快速迭代。

缺點：

- 完整釋出期漫長。
- 只適用於相容迭代的方式，如果重大版本不相容就沒辦法使用這種方式。

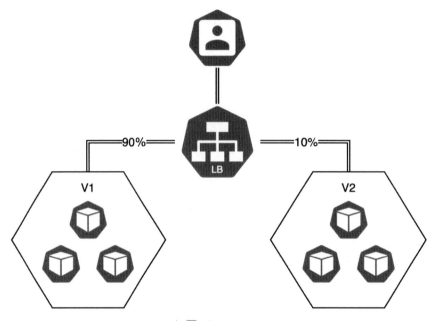

▲ 圖 13-4 canary

在金絲雀部署策略中，新版本的服務先在一小部分使用者中進行測試和調整。
這讓開發團隊能在問題擴大前，快速找到並修復錯誤，有效減少對整體用戶的
影響。這種策略特別適合對新版本有所疑慮或缺乏足夠測試的情況，它為新版
本的部署提供了一種保護機制。

然而，金絲雀部署並非萬能，例如在大版本不相容的情況下，或是需要長時間
才能完全釋出的情境，就無法使用這種方式。當然，每種部署策略都有其適用
的場景和限制，最終選擇哪種策略，取決於我們的應用特性和業務需求。

在實際操作中，部署策略的選擇需要綜合考慮各種因素，例如服務的規模、服
務的關鍵性、資源的可用性等。因此，理解各種部署策略的工作原理、優點和
缺點，對於選擇最適合自己的部署策略非常重要。

⯈ 13.5 A／B 測試（A／B Testing）

A／B 測試是一種與業務密切相關的技術，它基於統計數據而非僅僅是部署策略來指導業務決策。雖然這兩者是相關聯的，但也可以使用金絲雀部署來實現。

除了基於權重在版本之間進行流量控制之外，A／B 測試還可以基於一些其他參數（比如 Cookie、User Agent、地區等等）來精確定位給特定的用戶群，該技術廣泛用於測試一些功能特性的效果，並作為決策判斷的參考依據。

A／B 測試是線上同時執行多個不同版本的服務，這些服務更多的是使用者端的體驗不同，比如頁面布局、按鈕顏色，互動方式等，通常底層業務邏輯還是一樣的，也就是通常說的換湯不換藥。諸如 Google Analytics 等網站分析工具服務通常也可以搭配自家負載均衡器實現 A／B 測試。

優點：

- 多版本並行執行。
- 完全控制流量分布。

缺點：

- 需要更全面的負載均衡（通常由雲端服務實現）。
- 難以定位辨別（通常由雲端服務實現分散式追蹤）。

然而，A／B 測試並非無所不能。首先，實現 A／B 測試需要更全面的負載均衡，以控制不同版本的流量分布。此外，由於我們會有多個版本同時運行，因此可能會讓問題的定位變得較為困難。即使如此，透過精確的用戶群定位以及藉由資料驅動的決策，A／B 測試仍是一種強大的工具，能夠提供對用戶行為深入的理解，並導引我們改善產品。

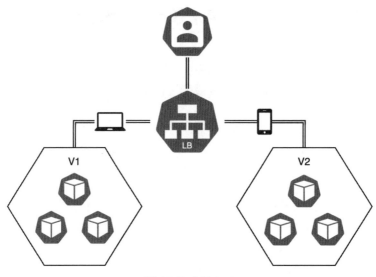

圖 13-5　A/B test

⫸ 13.6 影子部署（Shadow）

影子部署是指在原有版本旁完整運行新版本，並且將流入原有版本的請求，同時分發到新版本，得以實現在更新之前就模擬正式產品環境的運作情況，直到滿足上線條件後，才將進入點轉往新版本並關閉舊版本。

非常理想的流程，但背後所需要實現的技術門檻與成本相當的高，尤其是在特殊情況下，我們需要特別注意那些難以控制的情況。例如，一個下單的請求同時轉向新舊版本的服務，最終可能導致下單兩次的結果。

優點：

● 可以直接對正式環境流量進行效能測試而不影響使用者。
● 直到應用穩定且達到上線條件時才釋出。

缺點：

● 與藍綠部署一樣需要雙倍的資源請求。
● 設定複雜，容易出現預期外的情況。

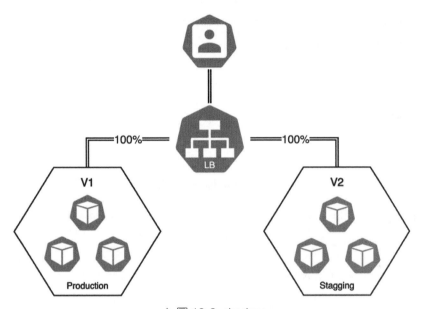

▲ 圖 13-6 shadow

實現影子部署的技術門檻和資源要求都相當高，需要精細的設計和設定以避免
預期外的問題。另一方面，由於需要運行與現有服務等量的新版本實例，影子
部署對資源的需求也是一大挑戰。

總之，雖然影子部署提供了強大的能力以保證新版本的穩定性，但同時也需要
有足夠的技術和資源承擔來實現它。實施影子部署需要進行仔細的評估，以確
定其是否適合特定的情況和需求。

筆者碎碎念

很高興這些曾經看似熟悉，卻說不出個所以然的策略，可以藉由這次的機會讓我一次將它們完整地介紹，但是只知道原理是遠遠不夠的。我們可是還沒發揮到 Kubernetes 在容器調度以及部署方面的強項呢，所以接下來的幾章我們即將回到歡樂的 Kubernetes 實戰環節。

參考資料

- 超詳細 GA 網站分析入門大全，看這篇就對了！
  https://www.webguide.nat.gov.tw/News_Content.aspx?n=531&s=2935

- 什麼是藍綠部署、金絲雀部署、滾動部署、紅黑部署、AB 測試？
  https://www.gushiciku.cn/pl/gUOs/zh-tw

- 5 Kubernetes Deployment Strategies: Roll Out Like the Pros
  https://spot.io/resources/kubernetes-autoscaling/5-kubernetes-deployment-strategies-roll-out-like-the-pros/?

- Six Strategies for Application Deployment
  https://thenewstack.io/deployment-strategies/

Kubernetes Deployment Strategies — 重建部署與滾動部署

在經過前面介紹的各種部署策略後，你是不是也跟我一樣對眼花撩亂的名稱豁然開朗，接下來我們就要一口氣練習其中兩種策略，分別是 Rolling Update 以及 Recreate，這兩者可以説是在 Kubernetes 中是最容易被實現的，因為它們正是資源物件 Deployment 中內建來汰換調度 Pod 的兩種選擇，話不多説就讓我們實際動手玩玩看吧。

14.1 重建部署（Recreate）

重建策略可以説是 Kubernetes 最容易實現的部署策略之一，因為它只要簡單地將 spec.strategy 設定為 Recreate 即可。定義為 Recreate 的 Deployment，會終止所有正在運行的實例，然後用較新的版本來重新建立它們，此技術如圖 14-1 所示，意味著服務的停機時間取決於應用程序的關閉和啟動持續時間，不建議在開發環境中使用。

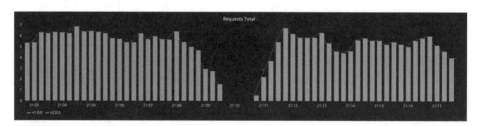

▲ 圖 14-1 recreate

建立 v1 版本應用服務：

範例檔 **recreate / app-v1.yaml**

```
1.  # recreate/app-v1.yaml
2.  apiVersion: apps/v1
3.  kind: Deployment
4.  metadata:
```

```
5.    name: foo-deployment
6.    labels:
7.      app: foo
8.  spec:
9.    replicas: 3
10.   strategy:
11.     type: Recreate
12.   selector:
13.     matchLabels:
14.       app: foo
15.   template:
16.     metadata:
17.       labels:
18.         app: foo
19.         version: v1
20.     spec:
21.       containers:
22.       - name: foo
23.           image: mikehsu0618/foo
24.           ports:
25.           - containerPort: 8080
```

這裡可以注意到我們利用 Deployment 原生支援的部署策略 spec.strategy.type: Recreate，代表我們可以預期當我們更新 Deployment 版本或內容時，其將會把所有 v1 版本的 Pod 完全關閉後，才會陸續啟動新版本。

讓我們運行起 v1 版本的服務：

```
1.  kubectl apply -f app-v1.yaml
2.  -------
3.  deployment.apps/foo-deployment-v1 created
```

接著我們將額外開啟一個終端機，並且使用 --watch 來觀察 Pod 之間的調度情形：

```
1.   # --watch 參數，終端機將會監聽所有來自 Pod 的狀態變化。
2.   kubectl get pods --watch
3.   --------
4.   NAME                              READY   STATUS    RESTARTS   AGE
5.   foo-deployment-8555547446-ld8fs   1/1     Running   0          8s
6.   foo-deployment-8555547446-nbt2k   1/1     Running   0          9s
7.   foo-deployment-8555547446-w6m5q   1/1     Running   0          13s
```

更新 v1 版本至 v2 版本：

範例檔 **recreate / app-v2.yaml**

```
1.   # recreate/app-v2.yaml
2.   apiVersion: apps/v1
3.   kind: Deployment
4.   metadata:
5.     name: foo-deployment
6.     labels:
7.       app: foo
8.   spec:
9.     replicas: 3
10.    strategy:
11.      type: Recreate
12.    selector:
13.      matchLabels:
14.        app: foo
15.    template:
16.      metadata:
17.        labels:
18.          app: foo
```

```
19.        version: v2
20.    spec:
21.      containers:
22.        - name: foo
23.          image: mikehsu0618/foo
24.          ports:
25.            - containerPort: 8080
```

直接把服務升級到 v2 版本：

```
1.  kubectl apply -f app-v2.yaml
2.  -------
3.  deployment.apps/foo-deployment-v1 created
```

接著回來查看剛剛 --watch 監聽的 Pod 情況：

```
1.  .....
2.  NAME                                READY  STATUS             RESTARTS  AGE
3.  foo-deployment-8555547446-ld8fs     1/1    Running            0         8s
4.  foo-deployment-8555547446-nbt2k     1/1    Running            0         9s
5.  foo-deployment-8555547446-w6m5q     1/1    Running            0         13s
6.  foo-deployment-8555547446-nbt2k     1/1    Terminating        0         21s
7.  foo-deployment-8555547446-ld8fs     1/1    Terminating        0         22s
8.  foo-deployment-8555547446-w6m5q     1/1    Terminating        0         22s
9.  foo-deployment-6cbc7db745-vgmmq     0/1    Pending            0         0s
10. foo-deployment-6cbc7db745-8n6c7     0/1    Pending            0         0s
11. foo-deployment-6cbc7db745-5t7p6     0/1    Pending            0         0s
12. foo-deployment-6cbc7db745-8n6c7     0/1    ContainerCreating  0         0s
13. foo-deployment-6cbc7db745-vgmmq     0/1    ContainerCreating  0         0s
14. foo-deployment-6cbc7db745-5t7p6     0/1    ContainerCreating  0         1s
15. foo-deployment-6cbc7db745-8n6c7     1/1    Running            0         6s
16. foo-deployment-6cbc7db745-5t7p6     1/1    Running            0         8s
17. foo-deployment-6cbc7db745-vgmmq     1/1    Running            0         12s
```

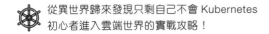

可以明確看到，所有的 Pod 同時被 Terminated 之後，才啟動更新後的
Deployment 的 v2 版本，沒錯就是這麼簡單優雅。

⚙ 14.2 滾動部署（Rolling Update）

滾動部署為 Kubernetes 中 Deployment 對所有 Pod 的預設部署策略，顧名思
義為在更新的過程中所有服務不會如同 Recreate 一樣極端地關閉所有服務，而
是如圖 14-2 般的小步快跑、細水長流般地平緩更新直到新版本就緒，如果沒有
特地宣告將會預設使用滾動更新調度 Pod，如需特地宣告，可以表示為 spec.
strategy.type: RollingUpdate， 或 是 設 定 strategy.rollingUpdate.{maxSurge/
maxUnavailable} 來控制服務更新的速率以確保穩定性。

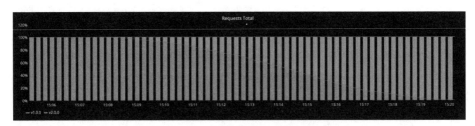

▲ 圖 14-2 rolling-update

建立 v1 版本應用服務：

範例檔 **rolling-update / app-v1.yaml**

```
1.  # rolling-update/app-v1.yaml
2.  apiVersion: apps/v1
3.  kind: Deployment
4.  metadata:
5.    name: foo-deployment
6.    labels:
7.      app: foo
```

```
8.  spec:
9.    replicas: 3
10.   strategy:
11.     type: RollingUpdate
12.     rollingUpdate:
13.       maxSurge: 1
14.       maxUnavailable: 0
15.   selector:
16.     matchLabels:
17.       app: foo
18.   template:
19.     metadata:
20.       labels:
21.         app: foo
22.         version: v1
23.     spec:
24.       containers:
25.       - name: foo
26.         image: mikehsu0618/foo
27.         ports:
28.           - containerPort: 8080
```

在 RollingUpdate 中多了一些參數可以來幫我們控制服務更新的速率：

- spec.strategy.rollingUpdate.maxSurge：此參數定義了在滾動更新過程中，可以超過我們原始應用 Pod 數量的最大值。這是一個非常重要的參數，可以控制我們在更新的過程中，允許同時有多少額外的 Pod 在運行。例如，如果設定為「10%」，則在滾動更新過程中，將會有最多額外 10% 的 Pod 被建立出來。這可以用來確保在更新過程中我們有足夠的資源來提供服務，不會因為更新導致服務無法滿足需求。

- spec.strategy.rollingUpdate.maxUnavailable：此參數定義了在滾動更新過程中，我們可以容忍的 Pod 不可用的最大數量或百分比。例如，如果設定為

「1」，則在任何時候都至少有一個 Pod 不可用。如果設定為「10%」，則在任何時候，最多 10% 的 Pod 允許為不可用。這個參數可以讓我們控制在更新過程中能夠容忍的服務中斷水平。

 NOTE

需要注意的是，這些設定在不同的使用場景和服務要求下，可能需要做不同的調整。例如，如果你的應用程式對於服務的可用性要求非常高，你可能需要設定一個較高的 maxSurge 和較低的 maxUnavailable。反之，如果你的應用程式可以容忍一定的服務中斷，而你希望能夠更快地完成更新，你可能需要設定一個較低的 maxSurge 和較高的 maxUnavailable。

讓我們運行其 v1 版本的服務：

```
1.  kubectl apply -f app-v1.yaml
2.  -------
3.  deployment.apps/foo-deployment configured
```

接著我們將額外開啟一個終端機並且使用 --watch 來觀察 Pod 之間的調度情形：

```
1.  kubectl get pods --watch
2.  --------
3.  NAME                              READY   STATUS    RESTARTS   AGE
4.  foo-deployment-8555547446-mrmb8   1/1     Running   0          7s
5.  foo-deployment-8555547446-vwtp4   1/1     Running   0          9s
6.  foo-deployment-8555547446-kg5f7   1/1     Running   0          12s
```

更新 v1 版本至 v2 版本：

範例檔 **rolling-update / app-v2.yaml**

```
1.  # rolling-update/app-v2.yaml
```

```
2.   apiVersion: apps/v1
3.   kind: Deployment
4.   metadata:
5.     name: foo-deployment
6.     labels:
7.        app: foo
8.   spec:
9.     replicas: 3
10.    strategy:
11.      type: RollingUpdate
12.      rollingUpdate:
13.        maxSurge: 1
14.        maxUnavailable: 0
15.    selector:
16.      matchLabels:
17.        app: foo
18.    template:
19.      metadata:
20.        labels:
21.          app: foo
22.          version: v2
23.      spec:
24.        containers:
25.        - name: foo
26.          image: mikehsu0618/foo
27.          ports:
28.            - containerPort: 8080
```

直接把服務升級到 v2 版本：

```
1.   kubectl apply -f app-v2.yaml
2.   -------
3.   deployment.apps/foo-deployment configured
```

接著回來查看剛剛 --watch 監聽的 Pod 情況：

```
1.   .....
2.   NAME                               READY   STATUS             RESTARTS     AGE
3.   foo-deployment-8555547446-mrmb8    1/1     Running            0              7s
4.   foo-deployment-8555547446-vwtp4    1/1     Running            0              9s
5.   foo-deployment-8555547446-kg5f7    1/1     Running            0             12s
6.   foo-deployment-6cbc7db745-vh8tn    0/1     Pending            0              0s
7.   foo-deployment-6cbc7db745-vh8tn    0/1     ContainerCreating  0              0s
8.   foo-deployment-6cbc7db745-vh8tn    1/1     Running            0              7s
9.   foo-deployment-8555547446-vwtp4    1/1     Terminating        0            104s
10.  foo-deployment-6cbc7db745-crhgt    0/1     Pending            0              0s
11.  foo-deployment-6cbc7db745-crhgt    0/1     ContainerCreating  0              0s
12.  foo-deployment-8555547446-vwtp4    0/1     Terminating        0            105s
13.  foo-deployment-6cbc7db745-crhgt    1/1     Running            0              6s
14.  foo-deployment-8555547446-kg5f7    1/1     Terminating        0            110s
15.  foo-deployment-6cbc7db745-5q5tf    0/1     Pending            0              1s
16.  foo-deployment-6cbc7db745-5q5tf    0/1     ContainerCreating  0              1s
17.  foo-deployment-8555547446-kg5f7    0/1     Terminating        0            111s
18.  foo-deployment-6cbc7db745-5q5tf    1/1     Running            0              5s
19.  foo-deployment-8555547446-mrmb8    1/1     Terminating        0            115s
```

如上我們可以觀察到 Deployment 同時的運行新服務以及停用舊服務，並且優雅地依照我們設定的速率完成更新。

恭喜各位我們一口氣就實現了兩種部署策略囉！

筆者碎碎念

介紹完了最基本的部署策略後，我們將要繼續往下前進到進階策略了（我好興奮啊），在這邊需要讓各位知道部署策略並不是某種特定不互相交集的部署技術，而是由基礎部署策略建立起更細膩的進階設定，且可以同時實現不只一種部署策略，就讓我們期待接下來的金絲雀部署的實戰演練吧。

Kubernetes Deployment Strategies — 金絲雀部署

在前面的章節我們單純使用了 Service 就實現出了藍綠部署，但你可能會心想：那說好用 Ingress 實現的方法呢？是不是我隨便講講別人就隨便信信，別急別急你們仔細看看現在不就來了嗎？沒錯！繼前面的 Service 操作後，本章要分享的是使用我們的路由守護神 Ingress 實現金絲雀部署的練習。

15.1 Nginx Ingress 金絲雀部署功能介紹

Ingress 基於七層的 HTTP 和 HTTPS 協議進行轉發，可以透過域名和路徑對訪問做到更細粒度的劃分。Ingress 作為 Kubernetes 叢集中一種獨立的資源，需要透過建立它來製定外部訪問流量的轉發規則，並透過 Ingress Controller 將其分配到一個或多個 Service 中。Ingress Controller 在不同供應商之間有著不同的實現方式，Kubernetes 官方維護的 Controller 為 Nginx Ingress，它支援透過設定註解（Annotations）來實現不同場景下的部署和測試。

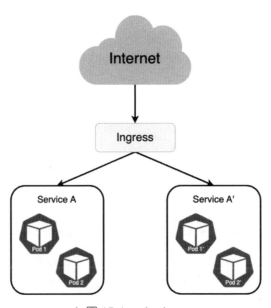

▲ 圖 15-1　nginx ingress

目前 Nginx Ingress 提供基於 Header、Cookie、權重三種外部流量切分策略，只需要在註解（Annotations）中寫入它提供的設定即可使用：

- nginx.ingress.kubernetes.io/canary：其值為 true 的話，將被視為 Canary Ingress，為以下設定進行流量切分，使兩個 Ingress 互相配合。
- nginx.ingress.kubernetes.io/canary-by-header-value：通知 Ingress 如有與 Header 設定值匹配的請求 Header，轉導流量到 Canary Ingress。
- nginx.ingress.kubernetes.io/canary-by-header-pattern：其運作方式與 canary-by-header-value 相同，並且支援正規表達式配對它，要注意的是當 canary-by-header-value 有被設定時，此註解的功能將會被忽略。
- nginx.ingress.kubernetes.io/canary-by-cookie：通知 Ingress 如有與 Cookie 設定值匹配的請求 Cookie，轉導流量到 Canary Ingress。如果將其值設定為 always，將會轉導所有流量。
- nginx.ingress.kubernetes.io/canary-weight：此數值預設為基於零到一百的整數，代表著宣告有多少百分比的流量將會被轉導到 Canary Ingress。
- 以上設定的優先級由高到低分別為：canary-by-header > canary-by-cookie > canary-weight。

▶ 15.2 金絲雀部署（Canary Deployment）

金絲雀部署與藍綠部署最大的不同是，它不是非黑即白的部署方式，而是介在於黑與白之間，能夠平滑過渡到下一個版本的方法。它能夠逐步地將修改推送給小部分的使用者，確定沒問題後才正式迭代到下一個版本，以降低直接引入新功能的風險。

圖 15-2 為金絲雀部署新舊版本更新過程中接收流量的示意圖：

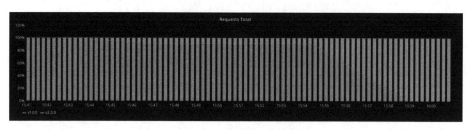

▲ 圖 15-2 canary deployment

▶ 15.3 使用金絲雀部署更新服務

本小節將會使用 Nginx Ingress 提供的 canary 功能來實現，藉由在註釋中指定需要被分流的權重比例。

大致實現方法可以分為以下步驟：

- 啟動一個原有的 v1 版本服務，並且使其與 Ingress 綁定成為唯一對外的正式版本。
- 啟動並且等待我們的 v2 版本完全就緒，此時新舊兩個版本處於同時存在的狀態。
- 加入 Canary Ingress 並設定預期分流到 v2 版本的請求權重。
- 直到確認 v2 版本有足夠條件取代 v1 版本後，將 Ingress 指向 v2 版本並且關閉 Canary Ingress。
- 確保成功終止舊的 v1 版本。

大致了解後就馬上來實現吧！

⯈ 15.4 實戰演練

範例檔 **app-v1.yaml**

```
1.   # app-v1.yaml
2.   apiVersion: apps/v1
3.   kind: Deployment
4.   metadata:
5.     name: foo-deployment
6.     labels:
7.       app: my-app
8.   spec:
9.     replicas: 1
10.    selector:
11.      matchLabels:
12.        app: my-app
13.        version: v1
14.    template:
15.      metadata:
16.        labels:
17.          app: my-app
18.          version: v1
19.      spec:
20.        containers:
21.          - name: foo
22.            image: mikehsu0618/foo
23.            ports:
24.              - containerPort: 8080
25.  ---
26.  apiVersion: v1
27.  kind: Service
28.  metadata:
```

```
29.    name: foo-service
30. spec:
31.    selector:
32.       app: my-app
33.       version: v1
34.    type: NodePort
35.    ports:
36.     - protocol: TCP
37.       port: 8080
38.       targetPort: 8080
```

範例檔 **app-v2.yaml**

```
1.  # app-v2.yaml
2.  apiVersion: apps/v1
3.  kind: Deployment
4.  metadata:
5.    name: bar-deployment
6.    labels:
7.      app: my-app
8.  spec:
9.    replicas: 1
10.   selector:
11.     matchLabels:
12.       app: my-app
13.       version: v2
14.   template:
15.     metadata:
16.       labels:
17.         app: my-app
18.         version: v2
19.     spec:
20.       containers:
```

```
21.            - name: bar
22.              image: mikehsu0618/bar
23.              ports:
24.                - containerPort: 8080
25. ---
26. apiVersion: v1
27. kind: Service
28. metadata:
29.   name: bar-service
30. spec:
31.   selector:
32.     app: my-app
33.     version: v2
34.   type: NodePort
35.   ports:
36.     - protocol: TCP
37.       port: 8080
38.       targetPort: 8080
```

範例檔 **ingress.yaml**

```
1.  # ingress.yaml
2.  apiVersion: networking.k8s.io/v1
3.  kind: Ingress
4.  metadata:
5.    name: my-ingress
6.  spec:
7.    ingressClassName: nginx
8.    defaultBackend:
9.      service:
10.       name: foo-service
11.       port:
12.         number: 8080
```

首先我們將運行 v1 版本並且啟用 Ingress 服務當作我們的 LoadBalancer：

```
1.  kubectl apply -f app-v1.yaml,ingress.yaml
2.  --------
3.  deployment.apps/foo-deployment created
4.  service/foo-service created
5.  ingress.networking.k8s.io/my-ingress created
```

查看 v1 版本服務以及 Ingress 狀態：

```
1.  kubectl get ingress
2.  --------
3.  NAME          CLASS   HOSTS   ADDRESS      PORTS   AGE
4.  my-ingress    nginx   *       localhost    80      2m14s
```

```
1.  kubectl get all
2.  --------
3.  NAME                                       READY   STATUS     RESTARTS   AGE
4.  pod/foo-deployment-68df868866-hjsdx        1/1     Running    0          82s
5.
6.  NAME               TYPE      CLUSTER-IP    EXTERNAL-IP   PORT(S)      AGE
7.  service/foo-service   NodePort   10.109.223.51 <none>    8080: 30256/TCP
    82s
8.  service/kubernetes    ClusterIP   10.96.0.1    <none>    443/TCP      32d
9.
10. NAME                             READY   UP-TO-DATE   AVAILABLE    AGE
11. deployment.apps/foo-deployment   1/1     1            1            82s
12.
13. NAME                                         DESIRED   CURRENT   READY   AGE
14. replicaset.apps/foo-deployment-68df868866    1         1         1       82s
```

現在我們可以發送一些請求確認一下是否 v1 版本為唯一對外的服務：

```
1.  for i in {1..10}; do curl localhost; echo; done
```

```
2.  -------
3.  {"data": "Hello foo"}
4.  {"data": "Hello foo"}
5.  {"data": "Hello foo"}
6.  {"data": "Hello foo"}
7.  {"data": "Hello foo"}
8.  {"data": "Hello foo"}
9.  {"data": "Hello foo"}
10. {"data": "Hello foo"}
11. {"data": "Hello foo"}
12. {"data": "Hello foo"}
```

接下來就把 v2 版本也完整啟動：

```
1.  kubectl apply -f app-v2.yaml
2.  ----------
3.  deployment.apps/bar-deployment created
4.  service/bar-service created
```

查看 v2 版本服務是否啟動完畢：

```
1.  kubectl get all
2.  ----------
3.  NAME                                     READY   STATUS    RESTARTS AGE
4.  pod/bar-deployment-7bbbff5c97-n7zhj      1/1     Running   0        52s
5.  pod/foo-deployment-68df868866-hjsdx      1/1     Running   0        14m
6.
7.  NAME                TYPE       CLUSTER-IP      EXTERNAL-IP  PORT(S)      AGE
8.  service/bar-service NodePort   10.97.149.224   <none>       8080: 32756/
    TCP    52s
9.  service/foo-service NodePort   10.109.223.51   <none>       8080: 30256/
    TCP    14m
10. service/kubernetes  ClusterIP  10.96.0.1       <none>       443/TCP      32d
11.
```

```
12. NAME                                    READY   UP-TO-DATE   AVAILABLE   AGE
13. deployment.apps/bar-deployment          1/1     1            1           52s
14. deployment.apps/foo-deployment          1/1     1            1           14m
15.
16. NAME                                       DESIRED   CURRENT   READY   AGE
17. replicaset.apps/bar-deployment-7bbbff5c97   1         1         1       52s
18. replicaset.apps/foo-deployment-68df868866   1         1         1       14m
```

確認目前依然是只開放 v1 版本接收請求：

```
1.  for i in {1..10}; do curl localhost; echo; done
2.  -------
3.  {"data": "Hello foo"}
4.  {"data": "Hello foo"}
5.  {"data": "Hello foo"}
6.  {"data": "Hello foo"}
7.  {"data": "Hello foo"}
8.  {"data": "Hello foo"}
9.  {"data": "Hello foo"}
10. {"data": "Hello foo"}
11. {"data": "Hello foo"}
12. {"data": "Hello foo"}
```

成功運行，接下來將要迎接主角 Canary Ingress 登場。

1. 加入 Canary Ingress 實現請求依權重分流到新舊版本：

範例檔 **canary-ingress.yaml**

```
1.  # canary-ingress.yaml
2.  apiVersion: networking.k8s.io/v1
3.  kind: Ingress
4.  metadata:
5.    annotations:
```

```
6.        nginx.ingress.kubernetes.io/canary: "true"
7.        nginx.ingress.kubernetes.io/canary-weight: "10"
8.      name: canary-ingress
9.   spec:
10.    ingressClassName: nginx
11.    defaultBackend:
12.      service:
13.        name: bar-service
14.        port:
15.          number: 8080
```

此處我們設定將 10% 比重的請求分流到 bar-service 這個 v2 版本的服務。

建立 Canary Ingress 資源：

```
1.   kubectl apply -f canary-ingress.yaml
2.   ---------
3.   ingress.networking.k8s.io/canary-ingress created
```

此時我們可以預期對 localhost 的請求中有百分之十會由 v2 版本接收：

```
1.   for i in {1..10}; do curl localhost; echo; done
2.   {"data": "Hello foo"}
3.   {"data": "Hello foo"}
4.   {"data": "Hello bar"} # 出現了！！
5.   {"data": "Hello foo"}
6.   {"data": "Hello foo"}
7.   {"data": "Hello foo"}
8.   {"data": "Hello foo"}
9.   {"data": "Hello foo"}
10.  {"data": "Hello foo"}
11.  {"data": "Hello foo"}
```

到此我們已經輕鬆實現金絲雀部署了。

接下來我們可以不斷調整比重，直到條件許可時，就將 Ingress 設定為新版本的 v2 並且刪除過渡使用的 Canary Ingress 以及 v1 版本資源即可。

修改 Ingress 設定為 v2 版本的 bar-service：

```
1.  kubectl patch ingress my-ingress -p '{"spec": {"defaultBackend":
    {"service": {"name": "bar-service"}}}}'
2.  --------
3.  ingress.networking.k8s.io/my-ingress configured
```

最後再讓我們的 Canary Ingress 安全退場：

```
1.  kubectl delete -f canary-ingress.yaml
2.  ---------
3.  ingress.networking.k8s.io/canary-ingress deleted
```

如此一來我們就算是完成一遍絲滑的金絲雀部署。

筆者碎碎念

我們終於完成了進階部署策略篇，也藉由了部署這個大觀念不斷重複加深 Deployment Service Pod Ingress 這幫好兄弟的使用方法，再次恭喜堅持走到這裡的讀者。老話一句，部署的方式並不侷限於任何特定形式，尤其在工作上我們有更多需要顧慮的 X 因子，各種花式技巧推陳出新都是為了應付某種特定業務情境，我們唯一能做到的就是穩紮穩打以不變應萬變，把基礎概念打好才是解決問題的根本。話說可以實現第七層負載均衡的 Ingress Controller，事實上是一門非常大的學問，而對 LoadBalancer 更深入了解一定也是學習 Kubernetes 的重要課題，希望日後也能有機會做一個更深入的探討。

參考資料

- 什麼是藍綠部署、金絲雀部署、滾動部署、紅黑部署、AB 測試？
  https://www.gushiciku.cn/pl/gUOs/zh-tw

- 管理資源
  https://kubernetes.io/zh-cn/docs/concepts/cluster-administration/
  manage-deployment/#canary-deployments

- ContainerSolutions/k8s-deployment-strategies
  https://github.com/ContainerSolutions/k8s-deployment-strategies/tree/
  master/canary

- K8S 學習筆記之 Kubernetes 部署策略詳解
  https://cloud.tencent.com/developer/article/1411271

- 使用 Nginx Ingress 實現灰度發佈和藍綠發佈
  http://dockone.io/article/2434773

- Configure a canary deployment
  https://docs.mirantis.com/mke/3.5/ops/deploy-apps-k8s/nginx-ingress/
  configure-canary-deployment.html

- NGINX Ingress Controller Annotations
  https://ithelp.ithome.com.tw/articles/nginx.ingress.kubernetes.io/canary-
  by-cookie

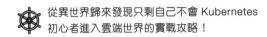
從異世界歸來發現只剩自己不會 Kubernetes
初心者進入雲端世界的實戰攻略！

Part 6
主題篇—Volume

相較之下 Docker Volume 好像遜色了點？

本篇深入探討了 Kubernetes 和 Docker 在資料持久化方面的核心功能 — Volume。我們將解析 Kubernetes 的 Volume 如何透過其豐富的類型和嚴謹的生命週期管理,優於 Docker Volume。讓我們一起來看看 Kubernetes 如何實現資料的無損重啟,並將這個概念推向了新的高度。

CHAPTER

16

Kubernetes Volume — Volume 是什麼？

相信有使用過 Docker 的同學們對 Volume 都不會太陌生，其主要功能是用來保存容器內的資料，此路徑資料夾內容將會與容器外的指定資料夾產生連接，即意味著這兩個資源是互通的，此後只要容器內的資料夾做任何存取，容器外的指定資料夾的內容也會跟著改變，讓我們可以在多個容器間共享一樣的資源，並且非常重要的是，當容器被刪除時，連結的資料夾以及裡面的檔案並不會被刪除，透過這個特性我們便能做到「刪除容器卻保留資料」的作用，確保容器重啟後能迅速還原資料和狀態。

16.1 那 Kubernetes 的 Volume 是什麼？

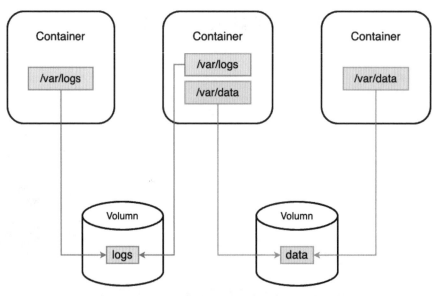

▲ 圖 16-1 Volume

會有如此標題是因為相較於 Docker，Kubernetes 的 Volume 擁有更豐富的類型以及更嚴謹的管理概念，並且 Kubernetes 在 Volume 中加入了生命週期的概念，使得我們擁有與 Pod 共生共滅的臨時卷（Ephemeral Volumes）和生命週期比 Pod 還長的儲存卷，並且在重啟容器期間不會丟失資料。

Volume 的核心就是一個目錄資料夾，其中可能存有資料，Pod 中的容器可以訪問該目錄中的資料。所採用的特定的卷類型將決定該目錄如何形成、使用何種介質保存資料以及目錄中存放的內容。

使用卷時，我們會在 .spec.volumes 欄位中設置這個卷以供 Pod 使用，並在 .spec.containers[*].volumeMounts 欄位中宣告卷在容器中的掛載位置。

16.2 Volume 類型

由於 Kubernetes 官方提供的類型實在過於精細並且迭代非常快速（每隔幾版就被棄用了），許多類型碰到的機會可以說是非常少，因此以下大致說明幾種最常見的 Volume 類型：

1. EmptyDir

當新增一個 Pod 時，Kubernetes 就會在新增一個 EmptyDir 的空白資料夾，讓此 Pod 中的所有容器都可以讀取這個特定的資料夾，常用於資料快取以及臨時儲存。

不過基於 EmptyDir 建構的 gitRepo Volume（現已被棄用）可以在 Pod 起始生命週期時，從對應的 gitRepo 中複製相對應檔案資料到底層的 EmptyDir 中，這使得它也可以具有一定意義上的持久性。

2. HostPath

HostPath 能將節點上的檔案或目錄掛載到你的 Pod 中，雖然這種需求不會太常遇到，但是它為了一些應用提供了更強大的後勤功能，例如在運行一個指定的 Pod 之前，先確認某 HostPath 下的檔案是否存在，以及應該以什麼方式存在。

NOTE

目前官方指明 HostPath 存在許多安全風險，最佳做法是盡可能避免使用 HostPath。當必須使用 HostPath 時，它的範圍應該要僅限於所需的檔案或目錄，並以只讀方式掛載。

3. Network FileSystem（NFS）

NFS 能將網路檔案掛載在你的 Pod 中。不同於 EmptyDir 會在刪除 Pod 的同時被刪除，這意味著它可以當作預先填充的資料，並且使這些資料在 Pod 之間共享。通常會搭配雲端儲存空間的服務使用。

4. ConfigMap

ConfigMap，從名稱便可推測它與設定相關。事實上，ConfigMap 主要用於存放設定檔。它可以作為我們常用的環境變數檔，或是用於資料庫初始化設定等部署相關的用途。

5. Secrets

從 Secrets 這名稱就可以知道，與 ConfigMap 相比，它主要用來存放敏感資料，例如使用者帳密或憑證。雖然它具備 ConfigMap 的所有功能，但也有其獨有的特性，例如將其內部資料進行 base64 編碼。

6. PV & PVC

PV（PersistentVolume）是叢集中的一塊儲存資源，可以由管理者事先設定，所以它們擁有獨立於任何使用 PV 的 Pod 的生命週期。

PVC（PersistentVolumeClaim）表達的是使用者對於儲存的請求，像是 Pod 會對 Node 的資源請求（CPU 或記憶體），PVC 同樣的也會請求並且消耗 PV 的限額。

▶ 16.3 不同 Volume 的生命週期

在 Kubernetes 中，不同類型的 Volume 擁有不同的生命週期，這對我們選擇適當的 Volume 類型非常關鍵。

1. EmptyDir

當 Pod 被建立在 Node 上時，就會建立 EmptyDir，並與 Pod 有著相同的生命週期。當 Pod 從節點中刪除時，EmptyDir 中的資料也將被永久刪除。

2. HostPath

HostPath 的生命週期與 Node 一致。也就是說，無論 Pod 的生命週期如何，只要節點存在，HostPath 中的資料就會保留下來。

3. Network FileSystem（NFS）

NFS 的生命週期超越 Pod 或 Node，資料會一直存在於 NFS 中，直到被明確地刪除。因此，當 Pod 或 Node 結束運行時，NFS 的資料仍將被保留。

4. ConfigMap

ConfigMap 的生命週期是獨立於使用它的 Pod。只有當它被明確刪除時，ConfigMap 及其資料才會被刪除。

5. Secrets

Secrets 的生命週期也跟 ConfigMap 一樣是獨立於使用它的 Pod。即使 Pod 被刪除，只要 Secrets 未被明確刪除，敏感資料仍將被保存。

6. PV & PVC

PVs 和 PVCs 的生命週期都超越了 Pod。即使 Pod 被刪除，只要 PVs 或 PVCs 未被明確刪除，相關的資料仍將被保存。值得注意的是，PV 和 PVC 之間的關係

會影響其生命週期，當 PVC 被刪除，如果其 reclaim policy 設定為「Delete」，
對應的 PV 也會被刪除。

筆者碎碎念

在 Kubernetes 中，如何設定資料的儲存與設置永遠都是一大課題跟學問，從我們
都稍微比較熟悉的 Volume 個人覺得是一個很好的切入點，接下來的幾章我們將會
開始實際進行操作練習，期望逐漸加深我們對於 Volume 的概念以及定位。

參考資料

- Kubernetes 文檔
  https://kubernetes.io/zh-cn/docs/

- Kubernetes 中的 emptyDir 儲存卷和節點儲存卷
  https://cloud.tencent.com/developer/article/1660415

- Kubernetes 那些事 — ConfigMap 與 Secrets
  https://medium.com/andy-blog/kubernetes-%E9%82%A3%E4%BA%9B%E4%BA%8B-configmap-%E8%88%87-secrets-5100606dd06c

Kubernetes Volume — EmptyDir

前面文章簡單介紹了 emptyDir，並且提到它與 Pod 的生命週期共生共滅，所以它通常被用於資料快取或者臨時儲存的場景，接下來就來實際操作練習一下吧。

17.1 EmptyDir Volume

emptyDir 可以簡單理解為在 Pod 運作時的一個臨時目錄，就像我們在 Docker 上使用的 Volume，而在 Pod 被移除時會被一併刪除，除了一些特殊場景通常我們都會將它用在一個 Pod 內的多個容器間的檔案共享，或做為容器資料的臨時快取儲存目錄等等。

emptyDir 儲存卷則定義於 .spec.volumes.emptyDir 嵌套欄位中，可用欄位主要包含兩個，具體如下：

● medium：此目錄所在儲存介質的類型，可取值為 default 或 Memory，預設為 default，表示使用節點的預設儲存介質；Memory 表示基於 RAM 的臨時檔案系統 tmpfs，空間受限於記憶體，但效能非常好，通常用於為容器中的應用提供快取空間。

● sizeLimit：當前儲存卷的空間限額，預設值為 nil，表示不限制；不過在 medium 欄位為 Memory 時，建議定義此限額。

17.2 實戰演練

1. 建立一個有多個容器的 Pod

範例檔 **emptydir-pod.yaml**

```
1.  # emptydir-pod.yaml
```

```
2.   apiVersion: v1
3.   kind: Pod
4.   metadata:
5.     name: emptydir-pod
6.   spec:
7.     volumes:
8.       - name: html
9.         emptyDir: {}
10.    containers:
11.      - name: nginx
12.        image: nginx: latest
13.        volumeMounts:
14.          - name: html
15.            mountPath: /usr/share/nginx/html
16.      - name: alpine
17.        image: alpine
18.        volumeMounts:
19.          - name: html
20.            mountPath: /html
21.        command: [ "/bin/sh", "-c" ]
22.        args: # 每十秒定時向 /html/index.html 寫入資料
23.          - while true; do
24.            echo $(hostname) $(date) >> /html/index.html;
25.            sleep 10;
26.            done
```

我們可以看到在 Pod 的設定檔中，我們啟用了 nginx 以及 alpine 並且掛載同一
個的 emptyDir 的共享目錄。

2. 建立 Pod

```
1.   kubectl apply -f emptydir-pod.yaml
2.   --------
```

```
3.  pod/emptydir-pod created
```

3. 查看 Pod 狀態

```
1.  kubectl describe pod/emptydir-pod
2.  --------
3.  …
4.      Mounts:
5.        /html from html (rw)
6.        /var/run/secrets/kubernetes.io/serviceaccount from kube-api-
    access-bmdwh (ro)
7.    Conditions:
8.      Type              Status
9.      Initialized       True
10.     Ready             True
11.     ContainersReady   True
12.     PodScheduled      True
13.   Volumes:
14.     html:
15.       Type:           EmptyDir (a temporary directory that shares a
    pod's lifetime)
16.       Medium:
17.       SizeLimit:  <unset>
```

可以從回傳結果中看到各個容器與 Volume 的掛載狀態。

4. 訪問 Pod 中的 Nginx

來確認一下 alpine 容器每隔 10 秒向 html/index.html 寫入訊息，而 Nginx 容器掛載的 emptyDir 是否同時也可以取得更新。

將 port 導出到本地的 localhost：

```
1.  kubectl port-forward pod/emptydir-pod 8080:80
```

```
2.  -------
3.  Forwarding from 127.0.0.1: 8080 -> 80pod 8080: 80
4.  Forwarding from [: : 1] : 8080 -> 80
```

使用 curl 查看回傳值：

```
1.  curl http://localhost:8080
2.  --------
3.  emptydir-pod Tue Jul 26 09: 05: 47 UTC 2022
4.  emptydir-pod Tue Jul 26 09: 05: 57 UTC 2022
5.  emptydir-pod Tue Jul 26 09: 06: 07 UTC 2022
6.  emptydir-pod Tue Jul 26 09: 06: 17 UTC 2022
7.  emptydir-pod Tue Jul 26 09: 06: 27 UTC 2022
```

順利取得由 alpine 容器產生的內容。

5. 進入容器查看

透過 -c 可以指定容器名稱進入指定容器。

```
1.  kubectl exec -it pods/emptydir-pod -c nginx -- sh
2.
3.  head -3 /usr/share/nginx/html/index.html
4.  --------
5.  emptydir-pod Tue Jul 26 09: 05: 47 UTC 2022
6.  emptydir-pod Tue Jul 26 09: 05: 57 UTC 2022
7.  emptydir-pod Tue Jul 26 09: 06: 07 UTC 2022
```

```
1.  kubectl exec -it pods/emptydir-pod -c alpine -- sh
2.
3.  head -3 /html/index.html
4.  --------
5.  emptydir-pod Tue Jul 26 09: 05: 47 UTC 2022
6.  emptydir-pod Tue Jul 26 09: 05: 57 UTC 2022
```

```
7.  emptydir-pod Tue Jul 26 09: 06: 07 UTC 2022
```

```
1.  ps aux
2.  --------
3.  PID    USER      TIME   COMMAND
4.    1 root       0 : 00 /bin/sh -c while true; do echo $(hostname)
    $(date) >> /html/index.html; sleep 10; done
5.   371 root      0: 00 sh
6.   395 root      0: 00 sleep 10
7.   396 root      0: 00 ps aux
```

6. 設定 Memory 作為高效能快取

範例檔 **emptydir-memory-pod.yaml**

```
1.  # emptydir-memory-pod.yaml
2.  apiVersion: v1
3.  kind: Pod
4.  metadata:
5.    name: emptydir-memory-pod
6.  spec:
7.    volumes:
8.      - name: html
9.        emptyDir:
10.         medium: Memory          # 指定使用記憶體儲存
11.         sizeLimit: 256Mi        # 限制記憶體大小
12.   containers:
13.     - name: nginx
14.       image: nginx: latest
15.       volumeMounts:
16.         - name: html
17.           mountPath: /usr/share/nginx/html
18.     - name: alpine
```

```
19.        image: alpine
20.        volumeMounts:
21.          - name: html
22.            mountPath: /html
23.        command: [ "/bin/sh", "-c" ]
24.        args:
25.          - while true; do
26.            echo $(hostname) $(date) >> /html/index.html;
27.            sleep 10;
28.            done
```

📠 筆者碎碎念

前面的簡單例子充分展現出 emptyDir 特有的定位以及簡單易懂的用法，但這些完全只是 Kubernetes Volumes 中多種類的其中之一，隨後幾章會來介紹其他常用的 Volume。

參考資料

- Kubernetes 中的 emptyDir 儲存卷和節點儲存卷
  https://cloud.tencent.com/developer/article/1660415

Kubernetes Volume — ConfigMap

在實際的產品開發環境中，我們往往會遇到需要在不同的環境之間切換部署，最常見的是開發環境（Development）和正式環境（Production）。這種切換通常涉及到不同的資料庫連線、使用的 Token、ApiKey 或初始化資料等設定變動。這些設定資訊的提取可以有效地降低程式碼的耦合度，只需修改相關設定檔，就能快速地建立所需的環境。而 Kubernetes 的 ConfigMap 給我們提供了一個非常方便的方式，讓我們能從最頂層輕鬆地注入這些設定。

▷ 18.1 ConfigMap 的特性

ConfigMap 基本上可以被視為是 Kubernetes 中的一種設定管理工具，它的核心功能在於存放設定檔，並在需要的時候提供這些設定檔的讀取。

以下是 ConfigMap 的一些主要特性：

● 靈活的儲存結構：一個 ConfigMap 物件可以存入一個或多個設定檔，這使得我們能夠靈活地管理和使用不同的設定。

● 降低程式碼耦合：透過 ConfigMap，我們可以在不改變原有程式碼的情況下，只需更換不同的 ConfigMap 即可實現不同環境的切換。

● 統一的設定管理：在 Kubernetes 中，所有的 ConfigMap 都可以在一個地方進行統一的查看和管理，這大大提升了設定管理的效率和便利性。

這些特性使得 ConfigMap 成為 Kubernetes 中不可或缺的一部分，並在實際的環境中發揮了巨大的作用。

⫸ 18.2 建立 ConfigMap

這裡簡單介紹建立一個 ConfigMap 的幾種方式：

1. 用指令匯入整個檔案

範例檔 **initdb.sql**

```
1.   # initdb.sql
2.   DROP TABLE IF EXISTS posts CASCADE;
3.
4.   CREATE TABLE posts
5.   (
6.       id              BIGSERIAL PRIMARY KEY,
7.       uuid            VARCHAR(36)  NOT NULL UNIQUE,
8.       user_id         NUMERIC      NOT NULL,
9.       title           VARCHAR(255) NOT NULL,
10.      content         TEXT         NOT NULL,
11.      comments_count  NUMERIC                 DEFAULT 0,
12.      created_at      TIMESTAMP    NOT NULL DEFAULT NOW(),
13.      updated_at      TIMESTAMP    NOT NULL DEFAULT NOW(),
14.      deleted_at      TIMESTAMP    NULL
15.  );
16.
17.  CREATE INDEX user_id_key ON posts (user_id);
18.
19.  COMMENT ON COLUMN posts.title IS '標題';
20.      COMMENT ON COLUMN posts.content IS '內容';
21.      COMMENT ON COLUMN posts.comments_count IS '評論數';
```

建立我們的 ConfigMap：

```
1.   kubectl create configmap pg-initsql --from-file=initdb.sql
2.   --------
3.   configmap/pg-initsql createdg-initsql
```

這裡我們利用 kubectl create 指令將整個檔案設定成一個 ConfigMap。

查看一下產生結果：

```
1.   kubectl describe configmap pg-initdb
2.   ---------
3.   Name:          pg-initdb
4.   Namespace:     default
5.   Labels:        <none>
6.   Annotations:   <none>
7.
8.   Data
9.   ====
10.  initdb.sql:
11.  ----
12.  DROP TABLE IF EXISTS posts CASCADE;
13.
14.  CREATE TABLE posts
15.  (
16.      id              BIGSERIAL PRIMARY KEY,
17.      uuid            VARCHAR(36)  NOT NULL UNIQUE,
18.      user_id         NUMERIC      NOT NULL,
19.      title           VARCHAR(255) NOT NULL,
20.      content         TEXT         NOT NULL,
21.      comments_count  NUMERIC                 DEFAULT 0,
22.      created_at      TIMESTAMP    NOT NULL DEFAULT NOW(),
23.      updated_at      TIMESTAMP    NOT NULL DEFAULT NOW(),
```

```
24.    deleted_at      TIMESTAMP      NULL
25. );
26.
27. CREATE INDEX user_id_key ON posts (user_id);
28.
29. COMMENT ON COLUMN posts.title IS '標題';
30.    COMMENT ON COLUMN posts.content IS '內容';
31.    COMMENT ON COLUMN posts.comments_count IS '評論數';
32.
33. BinaryData
34. ====
```

2. 使用指令建立 key-value 組合

```
1.   kubectl create configmap pg-connect \
2.   --from-literal=host=127.0.0.1 \
3.   --from-literal=port=5432
```

以上我們使用指令建立 host port 兩組 key-value。

查看一下結果：

```
1.   kubectl describe configmap pg-connect
2.   ----------
3.   Name:         pg-connect
4.   Namespace:    default
5.   Labels:       <none>
6.   Annotations:  <none>
7.
8.   Data
9.   ====
10.  host:
11.  ----
```

```
12. 127.0.0.1
13. port:
14. ----
15. 5432
16.
17. BinaryData
18. ====
```

3. 使用 yaml 檔建立設定檔

範例檔 **initdb-configmap.yaml**

```yaml
1.   # initdb-configmap.yaml
2.   apiVersion: v1
3.   kind: ConfigMap
4.   metadata:
5.     name: initdb-yaml
6.     labels:
7.       app: db
8.   data:
9.     initdb.sql: |
10.      DROP TABLE IF EXISTS posts CASCADE;
11.
12.      CREATE TABLE posts
13.      (
14.          id             BIGSERIAL PRIMARY KEY,
15.          uuid           VARCHAR(36)  NOT NULL UNIQUE,
16.          user_id        NUMERIC      NOT NULL,
17.          title          VARCHAR(255) NOT NULL,
18.          content        TEXT         NOT NULL,
19.          comments_count NUMERIC                 DEFAULT 0,
20.          created_at     TIMESTAMP    NOT NULL DEFAULT NOW(),
21.          updated_at     TIMESTAMP    NOT NULL DEFAULT NOW(),
```

```
22.       deleted_at        TIMESTAMP      NULL
23.    );
24.
25.    CREATE INDEX user_id_key ON posts (user_id);
26.
27.    COMMENT ON COLUMN posts.title IS '標題';
28.    COMMENT ON COLUMN posts.content IS '內容';
29.    COMMENT ON COLUMN posts.comments_count IS '評論數'
```

使用 yaml 檔建立我們的 ConfigMap：

```
1.   kubectl apply -f initdb-configmap.yaml
2.   ---------
3.   configmap/post-initdb-yaml created
```

查看一下結果：

```
1.   kubectl describe configmap initdb-yaml
2.   ---------
3.   Name:         initdb-yaml
4.   Namespace:    default
5.   Labels:       app=db
6.   Annotations:  <none>
7.
8.   Data
9.   ====
10.  initdb.sql:
11.  ----
12.  DROP TABLE IF EXISTS posts CASCADE;
13.
14.  CREATE TABLE posts
15.  (
16.      id            BIGSERIAL PRIMARY KEY,
```

```
17.    uuid            VARCHAR(36)  NOT NULL UNIQUE,
18.    user_id         NUMERIC      NOT NULL,
19.    title           VARCHAR(255) NOT NULL,
20.    content         TEXT         NOT NULL,
21.    comments_count NUMERIC                   DEFAULT 0,
22.    created_at      TIMESTAMP    NOT NULL DEFAULT NOW(),
23.    updated_at      TIMESTAMP    NOT NULL DEFAULT NOW(),
24.    deleted_at      TIMESTAMP    NULL
25. );
26.
27. CREATE INDEX user_id_key ON posts (user_id);
28.
29. COMMENT ON COLUMN posts.title IS '標題';
30. COMMENT ON COLUMN posts.content IS '內容';
31. COMMENT ON COLUMN posts.comments_count IS '評論數';
32.
33. BinaryData
34. ====
```

4. 使用 yaml 檔建立 key-value

範例檔 **initdb-kv.yaml**

```
1.  # initdb-kv.yaml
2.  apiVersion: v1
3.  kind: ConfigMap
4.  metadata:
5.    name: initdb-kv-yaml
6.    labels:
7.      app: db
8.  data:
9.    PG_USER: postgres
10.   PG_PASSWORD: postgres
```

查看一下結果：

```
1.  kubectl describe configmap initdb-kv-yaml
2.  --------
3.  Name:          initdb-kv-yaml initdb-kv-yaml
4.  Namespace:     default
5.  Labels:        app=db
6.  Annotations:   <none>
7.
8.  Data
9.  ====
10. PG_PASSWORD:
11. ----
12. postgres
13. PG_USER:
14. ----
15. postgres
16.
17. BinaryData
18. ====
```

⫸ 18.3 實戰演練

範例檔 **pg-pod.yaml**

```
1.  # pg-pod.yaml
2.  apiVersion: v1
3.  kind: Pod
4.  metadata:
5.    name: db
6.    labels:
```

```
7.       app: db
8.   spec:
9.     containers:
10.      - name: db
11.        image: postgres: 12.4-alpine
12.        env:
13.          # 使用 configmap 的 key-value 做為值傳入
14.          - name: POSTGRES_USER
15.            valueFrom:
16.              configMapKeyRef:
17.                name: initdb-kv-yaml
18.                key: PG_USER
19.          - name: POSTGRES_PASSWORD
20.            valueFrom:
21.              configMapKeyRef:
22.                name: initdb-kv-yaml
23.                key: PG_PASSWORD
24.          - name: PGDATA
25.            value: '/var/lib/postgresql/data/pgdata'
26.          - name: POSTGRES_DB
27.            value: 'posts'
28.        ports:
29.          - containerPort: 5432
30.        volumeMounts:
31.          # 使用 configmap 做為 file 當作初始化設定
32.          - mountPath: /docker-entrypoint-initdb.d
33.            name: initdb
34.     volumes:
35.       - name: initdb
36.         configMap:
37.           name: initdb
38. ---
```

```
39. apiVersion: v1
40. kind: ConfigMap
41. metadata:
42.   name: initdb
43.   labels:
44.     app: db
45. data:
46.   initdb.sql: |
47.     DROP TABLE IF EXISTS posts CASCADE;
48.
49.     CREATE TABLE posts
50.     (
51.         id             BIGSERIAL PRIMARY KEY,
52.         uuid           VARCHAR(36)  NOT NULL UNIQUE,
53.         user_id        NUMERIC      NOT NULL,
54.         title          VARCHAR(255) NOT NULL,
55.         content        TEXT         NOT NULL,
56.         comments_count NUMERIC               DEFAULT 0,
57.         created_at     TIMESTAMP    NOT NULL DEFAULT NOW(),
58.         updated_at     TIMESTAMP    NOT NULL DEFAULT NOW(),
59.         deleted_at     TIMESTAMP    NULL
60.     );
61.
62.     CREATE INDEX user_id_key ON posts (user_id);
63.
64.     COMMENT ON COLUMN posts.title IS '標題';
65.     COMMENT ON COLUMN posts.content IS '內容';
66.     COMMENT ON COLUMN posts.comments_count IS '評論數';
67. ---
68. apiVersion: v1
69. kind: ConfigMap
70. metadata:
```

```
71.    name: initdb-kv-yaml
72.    labels:
73.      app: db
74. data:
75.   PG_USER: PG_USER
76.   PG_PASSWORD: PG_USER
```

建立模擬實戰的 pg-pod：

```
1.   kubectl apply -f pg-pod.yaml
2.   -----------
3.   pod/db created
4.   configmap/initdb unchanged
5.   configmap/initdb-kv-yaml unchanged
```

在上面的設定中我們同時利用 ConfigMap 的檔案掛載以及 key-value 掛載，設定了我們的 postgres 初始化資料表以及使用者帳號密碼。

接下來我們來驗證一下是否設定都已成功注入到 postgres 中。

進入容器：

```
1.   kubectl exec -it db -- sh
2.
3.   ls
4.   ---------
5.   bin     etc     media       proc      sbin         tmp
6.   dev     home    mnt         root      srv          usr
7.   docker-entrypoint-initdb.d  lib   opt    run   sys     var
```

成功進入 postgres 即可使用 psq cli 操作。

檢查 initdb.sql 是否成功掛載：

```
1.   cat docker-entrypoint-initdb.d/initdb.sql
```

```
2.   ---------
3.   DROP TABLE IF EXISTS posts CASCADE;
4.
5.   CREATE TABLE posts
6.   (
7.       id              BIGSERIAL PRIMARY KEY,
8.       uuid            VARCHAR(36)  NOT NULL UNIQUE,
9.       user_id         NUMERIC      NOT NULL,
10.      title           VARCHAR(255) NOT NULL,
11.      content         TEXT         NOT NULL,
12.      comments_count NUMERIC                 DEFAULT 0,
13.      created_at      TIMESTAMP    NOT NULL DEFAULT NOW(),
14.      updated_at      TIMESTAMP    NOT NULL DEFAULT NOW(),
15.      deleted_at      TIMESTAMP    NULL
16. );
17.
18. CREATE INDEX user_id_key ON posts (user_id);
19.
20. COMMENT ON COLUMN posts.title IS '標題';
21. COMMENT ON COLUMN posts.content IS '內容';
22. COMMENT ON COLUMN posts.comments_count IS '評論數';
```

在這裡我們成功讀取掛載進去的 ConfigMap 檔案。

檢查 Role 是否被成功創立：

在剛剛的 postgres 容器中，使用 psql cli 登入：

```
1.   psql -U PG_USER -d posts
2.   -------
3.   psql (12.4)
4.   Type "help" for help.
5.
6.   posts=#
```

成功以我們設定的帳號名稱 PG_USER 以及資料庫 posts 登入。

檢查是否成功建立資料表 Posts：

```
1.  posts=# \d posts
2.                  Table "public.posts"
3.       Column    |    Type    | Collation | Nullable |       Default
4.   --------------+----------+----------+--------+------------------
5.   id | bigint |      | not null | nextval('posts_id_seq': : regclass)
6.   uuid     | character varying(36)    |        | not null |
7.   user_id  | numeric                  |        | not null |
8.   title    | character varying(255)   |        | not null |
9.   content      | text        |         | not null |
10.  comments_count| numeric    |         |          | 0
11.  created_at   | timestamp without time zone |  | not null | now()
12.  updated_at   | timestamp without time zone |  | not null | now()
13.  deleted_at    | timestamp without time zone |  |          |
14. Indexes:
15.     "posts_pkey" PRIMARY KEY, btree (id)
16.     "posts_uuid_key" UNIQUE CONSTRAINT, btree (uuid)
17.     "user_id_key" btree (user_id)
```

大功告成！

📠 **筆者碎碎念**

我們一起探索了 ConfigMap 的多種實用技巧，並實際進行了操作。儘管這只是 ConfigMap 功能的冰山一角，但這裡所提到的知識已經足夠應對大多數的應用場景。ConfigMap 的潛力和可能性遠超我們的想象，鼓勵大家深入研究並實踐官方文件中更進階的操作方式。

參考資料

- 配置 Pod 使用 ConfigMap
  https://kubernetes.io/zh-cn/docs/tasks/configure-pod-container/
  configure-pod-configmap/

- [Day 18] 高彈性部署 Application – ConfigMap
  https://ithelp.ithome.com.tw/articles/10196153

- Kubernetes 那些事 — ConfigMap 與 Secrets
  https://medium.com/andy-blog/kubernetes-%E9%82%A3%E4%BA
  %9B%E4%BA%8B-configmap-%E8%88%87-secrets-5100606dd06c

從異世界歸來發現只剩自己不會 Kubernetes
初心者進入雲端世界的實戰攻略！

Kubernetes Volume — Secret

前一章我們介紹了 Kubernetes 中的 ConfigMap，這是一個強大的工具，能讓我們解耦程式碼的複雜度，並統一管理設定檔。然而，ConfigMap 是以明碼形式儲存較不敏感的資訊，所以對於需要保護的敏感資料如 API Key 或金鑰等，ConfigMap 並不完全適用。為此，Kubernetes 提供了另一種解決方案——Secret。Secret 的使用方式類似於 ConfigMap，但它提供了額外一層保護，以防止敏感資訊被隨意暴露。然而，有趣的是，Secret 在某些方面可能並不像其名稱暗示的那樣「祕密」或「安全」。在接下來的內容中，我們將深入探討 Secret 的運作原理。

▶▶ 19.1 什麼是 Secret ？

Secret 在 Kubernetes 中扮演著與 ConfigMap 類似但又略有不同的角色。Secret 是 Kubernetes 提供的一種方法，用於儲存敏感資料，例如 API key、密碼等。實際上，Kubernetes 自己也使用這種機制來存放 Access Token，並對 API 存取權限進行控制，確保外部服務不能隨意操作。

大致上，有三種類型的 Secret：

- Service Account：這種 Secret 由 Kubernetes 自動建立並掛載到 Pod 中，主要用於存取 Kubernetes API。你可以在 /run/secret/kubernetes.io/serviceaccount 目錄中找到這種類型的 Secret。
- Opaque：這是以 base64 編碼的 Secret，用來儲存敏感資料，例如密碼、金鑰等。
- Docker-registry：如果映像檔儲存在私有的 Registry 中，我們需要使用這種類型的 Secret。

在 Kubernetes 中，常見的存取 Secret 的方式包括：

- 將 Secret 當作環境變數使用。
- 把 Secret 檔案掛載在 Pod 中的某個路徑下面。
- 在 Pod 中加入 docker-registry secret，這樣我們在拉取私人倉庫的映像檔時就不需要每次都先進行 docker login。簡單來説，docker-registry secret 就是用來儲存 docker login 的帳號密碼，讓 Kubernetes 可以自動登入以順利運作。

關於 Secret 如何實現保護敏感資料，以及在 Kubernetes 中使用 Secret 需要注意的事項：

- 關於 Secret 的加密：雖然 Secret 對敏感資料進行了基本的保護（例如 base64 編碼），但是它們在 etcd 中仍以明文形式儲存。如果需要更強的安全保護，可以考慮使用額外的加密解決方案，例如 Kubernetes 的 Encryption-at-Rest 功能。
- 使用 Secret 的注意事項：儘管 Secret 可以保護敏感資料，但是我們也需要適當地管理和保護 Secret 自身。例如，我們應避免在日誌或錯誤訊息中暴露 Secret 的資訊，並且需要限制和監控對 Secret 的存取權限。
- Secret 與 ConfigMap 的最佳使用方式：雖然 Secret 和 ConfigMap 在功能上相似，但是我們應該明確地區分何時使用何者。一般來説，對於非敏感資訊，例如應用設定，我們可以使用 ConfigMap。對於需要保護的敏感資訊，例如密碼、金鑰等，則應該使用 Secret。

▶ 19.2 建立 Secret

在建立 Secret 的值時，我們都要必須先使用 base64 進行編碼，而 Kubernetes 在我們正確掛載後會自動幫我們解碼回原本的值。

將 Secret 轉換為 base64。首先我們可以使用內建語法取得經 base64 編碼過的
字串：

```
1.  echo -n 'my-account' | base64
2.  echo -n 'my-password' | base64
3.  ----------
4.  bXktYWNjb3VudA==
5.  bXktcGFzc3dvcmQ=
```

於是我們獲得了以上的 bXktYWNjb3VudA== bXktcGFzc3dvcmQ= 做為加密過
的字串。

接著建立 Secret 設定檔：

範例檔 **test-secret.yaml**

```
1.  # test-secret.yaml
2.  apiVersion: v1
3.  kind: Secret
4.  metadata:
5.    name: test-secret
6.  data:
7.    username: bXktYWNjb3VudA==
8.    password: bXktcGFzc3dvcmQ=
```

建立 Secret：

```
1.  kubectl apply -f test-secret.yaml
2.  ----------
3.  secret/test-secret created
```

查看一下結果：

```
1.  kubectl get secret test-secret
2.  ----------
```

```
3.   NAME             TYPE      DATA    AGE
4.   test-secret      Opaque    2       15s
```

成功建立 Secret 資源，接著讓我們查看更詳細的細節：

```
1.   kubectl describe secret test-secret
2.   ----------
3.   Name:          test-secret test-secret
4.   Namespace:     default
5.   Labels:        <none>
6.   Annotations:   <none>
7.
8.   Type:  Opaque
9.
10.  Data
11.  ====
12.  password:  11 bytes
13.  username:  10 bytes
```

可以在 Data 內看到 Secret 的 key 值和 base64 加密後的大小。

而我們可以只使用 kubectl 建立：

```
1.   kubectl create secret generic test-secret --from-literal=
     'username=my-account' --from-literal='password=my-password'
```

這裡一樣使用了 create 當作建立資源的宣告指令，並且還需要加上 generic 這個 subcommand，表示我們要使用本地資源或者是 key-value 建立 Secret。

⊪ 19.3 實際應用 Secret

讓我們利用前面建立的 Secret 來實際測試：

範例檔 **secret-test-pod.yaml**

```
1.  # secret-test-pod.yaml
2.  apiVersion: v1
3.  kind: Pod
4.  metadata:
5.    name: secret-test-pod
6.  spec:
7.    containers:
8.      - name: test-container
9.        image: nginx
10.       volumeMounts:
11.         - name: secret-volume
12.           mountPath: /etc/secret-volume
13.   volumes:
14.     - name: secret-volume
15.       secret:
16.         secretName: test-secret
```

建立 Pod 並查看：

```
1.  kubectl apply -f secret-test-pod.yaml
2.  -----------
3.  pod/secret-test-pod created
```

```
1.  kubectl get pod secret-test-pod
2.  ------------
3.  NAME              READY   STATUS    RESTARTS   AGE
4.  secret-test-pod   1/1     Running   0          57s
```

進入 Pod 中查看：

```
1.  kubectl exec -it secret-test-pod -- sh
```

在容器中印出 Secret 內容：

```
1.  ls /etc/secret-volume
2.  ------------
3.  password  username
4.  cat /etc/secret-volume/username
5.  ------------
6.  my-account
7.
8.  cat /etc/secret-volume/password
9.  ------------
10. my-password
```

成功印出 base64 解碼後的數值。

▶▶ 19.4 聊聊關於 Secret 看起來並不那麼安全這件事

在我們過去的探討中，我們認識到 Kubernetes 的 Secret 能有效地儲存和管理敏感資訊。但是，當我們仔細研究其實現方式，發現並不像我們想像的那樣安全。

首先，Secret 的內容經過了 base64 編碼，但這基本上等同於明文。此外，只要擁有相應的權限，任何人都可以輕鬆地查看 Secret 的原始內容。因此，原生的 Secret 不適合在對安全性要求較高的場景中使用，例如大型公司或對 RBAC（角色權限）需求較高的場景。

那麼，我們應該如何解決這個問題呢？一個可能的方案是使用 etcd 加密、API Server 的權限限制以及強化 Node 的權限管理和系統安全性。但是，這些方案需要一起實施，而且實施成本相對較高。

幸好，現在有許多解決方案可供我們選擇，例如 AWS Key Management Service 或 Google Cloud KMS 等。這些工具都提供了更強大的密鑰管理功能，可以幫助我們更有效地保護敏感資料。

然而，選擇合適的工具只是我們面臨的挑戰之一。我們還需要考慮如何符合各種安全性和合規性要求，例如，某些法規可能要求我們對某些資料進行特定的保護措施。在後面的章節中，我們將會深入探討這些主題，包括密鑰管理的最佳實踐，以及如何在 Kubernetes 中實現這些實踐。同時，我也會介紹一些關於使用 Kubernetes Secret 和 ConfigMap 的實際案例，以更好地理解這些工具在實際應用中的優點與缺點。

參考資料

- 使用 Secret 安全地分發憑證
  https://kubernetes.io/zh-cn/docs/tasks/inject-data-application/distribute-credentials-secure/

Kubernetes Volume
— PV & PVC

PV 與 PVC（以下簡稱 PV & PVC）的觀念常與如何使服務保持 Stateful 的設定相關。它們之所以重要，是因為 PV 的生命週期是獨立於 Pod 的。這意味著，無論我們何時終止或擴大 Pod，我們都可以持續保有原有的資料。

在實務上，這可以幫助我們建立一個資料持久化服務，例如一個資料庫或檔案儲存服務，即使服務的一部分（例如單個 Pod）發生故障，我們也能確保資料的安全和可用性。此外，資料儲存的位置已經從原本的 Pod 中轉移到了更為穩定且可靠的儲存空間，例如 NFS（Network File System）、叢集中的 Node 或雲端儲存服務，這讓我們的資料享有獨立的生命週期，並進一步提高了資料的持久性。

20.1 Storage Class

StorageClass 提供了一種描述和分類 PV 的方式。這是一種動態儲存設定的方法，允許叢集管理員定義儲存「類型」。當 PVC 被建立時，如果沒有靜態的 PV 可以滿足它的需求，Kubernetes 就會根據指定的 StorageClass 自動建立一個新的 PV。

以下是一個 StorageClass 的範例設定檔。這個範例展示了如何在 Google Kubernetes Engine（GKE）中建立一個 StorageClass。這個 StorageClass 的 provisioner 設定為 Google Persistent Disk，並且設定為「Standard」type，讓該 StorageClass 的 PV 使用標準的永久性磁碟：

```
1.  apiVersion: storage.k8s.io/v1
2.  kind: StorageClass
3.  metadata:
4.    name: standard
5.  provisioner: kubernetes.io/gce-pd
6.  parameters:
7.    type: pd-standard
```

```
8.  reclaimPolicy: Retain
```

StorageClass 的具體設定可能會因 Kubernetes 的運行環境（例如雲端平台或 On-Premise）以及你的特定需求（例如效能、可用性、價格等）而異。

每一種 provisioner 都有自己的參數，你可以設定這些參數來客製化你的儲存需求。在前面範例中，我們使用了 type 參數來指定我們希望使用的 Google Persistent Disk 的類型。

Persistent Volume 的回收政策（Reclaim Policy）是一個關鍵的設定，它定義了 Persistent Volume Claim（PVC）從 Persistent Volume（PV）中釋放後，如何處理該 PV。具體來說，Kubernetes 提供了三種回收政策：

- Retain：這個政策將保留 PV 和其所有的資料，即使 PVC 已經釋放。此時，PV 會變為「Released」狀態，但並不會被其他 PVC 再次使用，除非該 PV 已經被手動清理。
- Delete：當 PVC 釋放後，這個政策會自動刪除 PV 和其關聯的儲存，並從叢集中移除 PV。
- Recycle：這個政策已經被 Kubernetes 廢棄，原先的目的是在 PVC 釋放後對 PV 進行基本的資料清理，然後可以重新用於其他 PVC。

對於每個建立的 PV，根據需求選擇合適的回收政策是非常重要的，這可以幫助我們更好地控制 PV 的生命週期，並根據需要確保資料的保存或刪除。

⠿ 20.2 Persistent Volumes（PV）

Kubernetes 利用 PV 提供一個抽象的儲存空間，並且 PV 能以動態和靜態的被提供，可以簡單理解為當 PV 是預先被宣告，隨後被 PVC 取用的話就是一種靜態產生；而如果 PVC 中有指定 StorageClass 的種類時，Kubernetes 將會動態地為我們產生 PV。

當一位系統管理者宣告了一塊儲存空間後，而系統的使用者就能以 PVC 來請求
此儲存空間，形成圖 20-1 的關係：

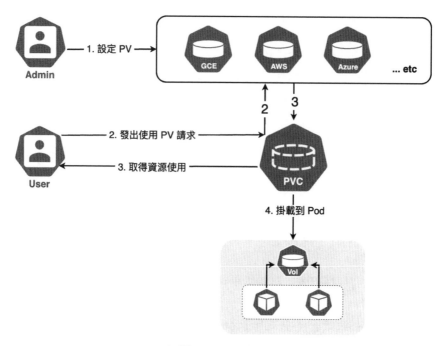

▲ 圖 20-1　PV relation

20.3 Persistent Volume Claims（PVC）

在 PVC 中表達的是使用者對儲存的請求，相較於 Pod 可以請求特定數量的
CPU 和記憶體資源，PVC 也可以請求儲存空間大小以及設定訪問模式（例如：
ReadWriteOnce、ReadOnlyMany 或 ReadWriteMany），而 PVC 的資源請求成
立後，將會去不斷找尋符合條件的 PV，直到找到符合條件的資源並且將兩者綁
定，如果找不到匹配的 PV 時，PVC 將會無限期處於未綁定狀態（Pending），
直到出現匹配的 PV 加入。

實際範例：

範例檔 **pvc-demo.yaml**

```
1.   # pvc-demo.yaml
2.   kind: PersistentVolumeClaim
3.   apiVersion: v1
4.   metadata:
5.     name: pvc-demo
6.   spec:
7.     accessModes:
8.       - ReadWriteOnce
9.     storageClassName: hostpath
10.    resources:
11.      requests:
12.        storage: 1Gi
```

以上我們使用 PVC 動態產生 PV 並且綁定兩者，讓我們來瞭解其中設定的差異：

1. 訪問模式（spec.accessModes）

- ReadWriteOnce：允許單一節點以讀寫方式掛載卷。在一個節點中，多個 Pod 都可以共享訪問這個卷。這種模式對於大多數應用來說是足夠的，但在多節點的環境下可能會遇到限制。

- ReadOnlyMany：允許多個節點以只讀方式掛載卷。對於一些需要共享相同資料但不需要寫入的應用來說，這是一個很好的選擇。

- ReadWriteMany：允許被多個節點以讀寫方式掛載。這在需要多個 Pod 同時讀寫同一個 PV 的場景下非常有用。例如，分散式應用或高併發資料庫。

- ReadWriteOncePod：一種新的訪問模式，只允許單個 Pod 以讀寫方式掛載卷，這對於需要高度隔離和保護的資料是一個很好的選擇。

訪問模式是 PV & PVC 中蠻值得注意的點，因為本書主要都是使用「單節點」的 Kubernetes（docker-desktop）叢集，並不會遇到不同節點中的 Pod 掛載 PV

的情況，而現實的生產環境裡，我們使用的 Google GKE 或 AWS ELK 通常都是多節點的情況，這時不同節點之間 Pod 以及服務就必須指定 ReadOnlyMany 或 ReadWriteMany，才能順利取得共享資源，加上支援這兩個類型的 Provisioner 選擇不多，其中多半是使用雲端的 NFS 服務實現。

即使是相同服務的 Pod，也不一定會被 Kubernetes 的資源分配器分配到同一個節點上，除非使用 NodeSelector 或 Affinity/Anti-Affinity 等設定。

2. 儲存類型（spec.storageClassName）

儲存類型（spec.storageClassName）的概念在 Kubernetes 中扮演著重要的角色。簡單來説，StorageClass 為 Kubernetes 提供了一種機制來描述和分類不同的儲存選項。當一個 PersistentVolumeClaim（PVC）被建立，如果沒有指定特定的 PersistentVolume（PV），Kubernetes 就會根據 PVC 所指定的 StorageClass 自動建立一個 PV。這個過程被稱為動態儲存設定，它可以大大提高儲資源管理的靈活性和自動化程度，而在某些需要更精確控制儲存資源分配的場景，或者在高度規範化的環境中，你可能會選擇停用動態儲存設定，並由系統管理員手動建立並綁定一個 PV。

舉例來説，在 docker-desktop 環境中，預設的 StorageClass 是「hostpath」，這意味著資料將直接儲存在單一節點的本地檔案系統中，這對於小型應用和測試環境來説或許適用，但在大規模和生產環境中，你可能需要使用更強大、更持久的儲存解決方案，例如雲端儲存服務或者分散式檔案系統。

值得一提的是，如果 PVC 沒有特別指定一個 StorageClass，Kubernetes 將會使用預設的 StorageClass。而如果你在 PVC 中設置了 StorageClass 為空字串 ""，這將停用動態儲存設定，這時你將需要手動建立並綁定一個 PV。

理解 StorageClass 不僅有助於我們理解 Kubernetes 的儲存管理方式，更可以讓我們根據實際需求選擇最適合的儲存方案。

⫸ 20.4 實戰演練

現在就讓我們使用 PVC 搭起兩個服務中間共用的儲存空間，模擬實際上多個服務對同一份資料的讀寫操作。

首先來建立我們的 PVC：

```
1.   kubectl apply -f ./pvc-demo.yaml
2.   ----------
3.   persistentvolumeclaim/pvc-demo created
```

查看結果：

```
1.   kubectl get pv
2.   ----------
3.   NAME                                        CAPACITY    ACCESS MODES
     RECLAIM POLICY    STATUS    CLAIM            STORAGECLASS    REASON    AGE
4.   pvc-c197c285-d314-43db-8cbc-6f912d8a9680    1Gi              RWO
     Delete            Bound     default/pvc-demo  hostpath                 2s
```

```
1.   kubectl get pvc
2.   ----------
3.   NAME           STATUS    VOLUME
     CAPACITY    ACCESS MODES    STORAGECLASS    AGE
4.   pvc-demo       Bound     pvc-c197c285-d314-43db-8cbc-6f912d8a9680    1Gi
     RWO                        hostpath         11s
```

可以看到 PV & PVC 都已經成功的綁定（Bound）並且設定相關的資源。

在 Kubernetes 系統中，PersistentVolume（PV）可能會存在四種不同的狀態（Status），每種狀態代表了 PV 在其生命週期中的不同階段：

● Available：此狀態表明 PV 已經建立並準備好被 PersistentVolumeClaim（PVC）使用。

- Bound：在此狀態下，PV 已經被特定的 PVC 綁定並正在使用中。
- Released：此狀態表示與 PV 綁定的 PVC 已被刪除，但 PV 本身尚未被回收。
- Failed：如果 PV 的回收過程中出現問題，則 PV 將會進入此狀態。

這四種狀態代表了 PV 從被建立到被刪除的完整生命週期，對於理解和管理
Kubernetes 的儲存系統非常重要。

建立兩個相關服務：

範例檔 **nginx-pod.yaml**

```
1.   # nginx-pod.yaml
2.   apiVersion: v1
3.   kind: Pod
4.   metadata:
5.     name: nginx-pod
6.   spec:
7.     containers:
8.     - name: nginx
9.       image: nginx: latest
10.    volumeMounts:
11.    - name: html
12.          mountPath: /usr/share/nginx/html
13.    volumes:
14.     - name: html
15.        persistentVolumeClaim:
16.          claimName: pvc-demo
17.          readOnly: false
```

範例檔 **apline-pod.yaml**

```
1.   # alpine-pod.yaml
2.   apiVersion: v1
```

```
3.   kind: Pod
4.   metadata:
5.     name: alpine-pod
6.   spec:
7.     containers:
8.       - name: alpine
9.         image: alpine
10.        command: [ "/bin/sh", "-c" ]
11.        args: # 每十秒定時向 /html/index.html 寫入資料
12.          - while true; do
13.            echo $(hostname) $(date) >> /html/index.html;
14.            sleep 10;
15.            done
16.           volumeMounts:
17.          - name: html
18.            mountPath: /html
19.      volumes:
20.        - name: html
21.          persistentVolumeClaim:
22.            claimName: pvc-demo
23.            readOnly: false
```

將前面兩個掛在 PVC 的 Pod 運行起來：

```
1.   kubectl apply -f ./nginx-pod.yaml,./alpine-pod.yaml
2.   -----------
3.   pod/alpine-pod configured
4.   pod/nginx-pod configured
```

訪問 Pod 中的 Nginx。

接著來確認一下 alpine 容器每隔 10 秒向 html/index.html 寫入訊息，而 Nginx
容器掛載的 PVC 是否同時也可以取得更新。

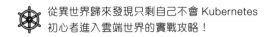

將 port 導出到本地的 localhost：

```
1.   kubectl port-forward pod/nginx-pod 8080:80
2.   -------
3.   Forwarding from 127.0.0.1: 8080 -> 80pod 8080: 80
4.   Forwarding from [: : 1] : 8080 -> 80
```

使用 curl 查看回傳值：

```
1.   curl http://localhost:8080
2.   --------
3.   alpine-pod Sat Aug 6 03: 31: 50 UTC 2022
4.   alpine-pod Sat Aug 6 03: 32: 00 UTC 2022
5.   alpine-pod Sat Aug 6 03: 32: 10 UTC 2022
6.   alpine-pod Sat Aug 6 03: 32: 20 UTC 2022
7.   alpine-pod Sat Aug 6 03: 32: 30 UTC 2022
```

順利取得由 alpine 容器產生的內容。

筆者碎碎念

PV & PVC 的觀念很廣，牽涉的範圍也很宏觀，需要對 Kubernetes 以及服務的運作場景有不少的理解。筆者認真鑽研了 PV & PVC 的觀念後，剛好遇到公司有使用需求。只能說保持資料的持久性這方面真的是一塊大議題，在 StackOverflow 中有網友分享道：Kubernetes 的精神更適合拿用來做 Stateless 的微服務，並且將需要持久性的資料抽象出來做為一個獨立服務並開放 API 以供其他服務取得。

在最近的工作上也有更深的一層的認同，筆者為了讓 Google GKE 上的多節點叢集實現共享持久性資料，就已經把 Google 提供的 Kubernetes Provision storage 服務都實作過一次，分別是建立在 Google FileStore 的雲端 NFS 服務（實現了 ReadWriteMany），以及建立在 Google Compute Engine Disk 的雲端硬碟服務（實現了 ReadWriteOnce），兩種服務也對應了不同的情況，之後有機會可以再來分享。

參考資料

- Using the Compute Engine persistent disk CSI Driver
  https://cloud.google.com/kubernetes-engine/docs/how-to/persistent-volumes/gce-pd-csi-driver

- Kuberentes 持久卷
  https://kubernetes.io/zh-cn/docs/concepts/storage/persistent-volumes/#access-modes

- Day 15 - 別再遺失資料了：Volume (2)
  https://ithelp.ithome.com.tw/articles/10193550

Part 7
主題篇—Resources

資源監控一定是全新的世界

曾經我們照著教學 Run 起了服務後，就覺得自己無堅不摧。殊不知接踵而來的維運、調校、監控才是另一個考驗的開始。

Kubernetes Resources — Resource

在一般的主機環境下，我們都知道 CPU 滿載時，很可能會帶來延遲以及鎖定，而 Memory 滿載（OOM）時，通常會伴隨著 Restart 以及啟動失敗。Kubernetes 為我們帶來了 Resources 中的 Request 以及 Limit 的概念，使我們可以在物理主機中實現抽象的資源隔離，達到資源利用率提高以及發揮容器彈性的特色。

21.1 Resource 是什麼？

為了使 Kubernetes 叢集可以更好的調度，在 Kubernetes 中的每個 Container 都可以設定以下兩種參數：

```
1.   resources:
2.       requests:
3.           cpu: 50m
4.           memory: 50Mi
5.       limits:
6.           cpu: 100m
7.           memory: 100Mi
```

- Requests：一個容器的最小啟動資源，代表 Kubernetes 叢集需要將它調度到滿足這個資源單位（CPU、Memory）的節點上。如果 Request 過低，可能會出現節點實際上並無足夠資源以啟動該容器的情況，反之如果 Request 過高，則會出現資源使用率過低，並阻止其他容器在該節點上進行調度，導致需要開啟額外的節點才可以正常運作。
- Limits：一個容器的最大可用資源，可以用來預防容器使用過多的資源，造成其他服務的資源短缺，以及容器中導致資源耗盡不可預期的問題。

簡單來說，Request 對於資源利用率以及調度更為重要，而 Limit 則注重於系統的穩定。

⊪ 21.2 Request 和 Limit 關係

容器所宣告的 Request 應該大於等於 0 並且不超過節點的可分配容量。這條規
則可用以下公式總結：

```
1.   0 <= request <= Node Allocatable
```

而 Limit 則應該大於等於 Request 並且其值無上限：

```
1.   request <= limit <= Infinity
```

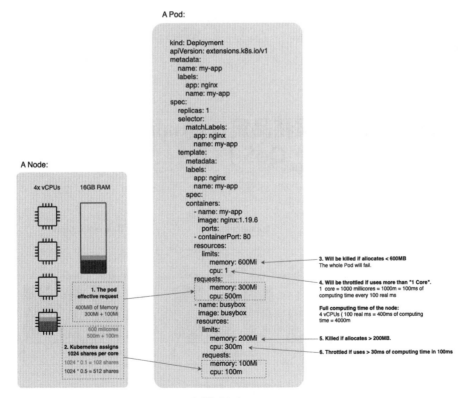

▲ 圖 21-1 resource

Kubernetes 將底層處理器架構抽象為了計算資源，將它們按照需求暴露為原始值或基本單位。

1. CPU：對於 CPU 資源來説，這些基本單位是基於核心（Core）的，而一個 CPU 則相當於：

   - 一個 AWS vCPU
   - 一個 GCP Core
   - 一個 Azure vCore
   - Intel 處理器上一個 Hyperthread（處理器要支援 Hyperthreading）

2. Memory：對於記憶體來説，則是基於位元組的。記憶體資源可以使用單純的數值或帶有後綴（E、P、T、G、M、K）的定點整數表示，也就是我們常見的單位。

21.3 Pod 的服務品質（Quality of Service，QoS）

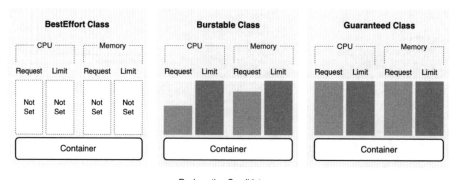

▲ 圖 21-2　QoS

Kubernetes 建立 Pod 時就會依照設定的 Request / Limit 給它指定了以下一種 QoS 分類：

1. Guaranteed

當一個 Pod 內的每個容器，其 request.memory 等於 limit.memory 且 request. cpu 等於 limit.cpu 時，這個 Pod 被認為是 Guaranteed。

```
1.   apiVersion: v1
2.   kind: Pod
3.   metadata:
4.     name: qos-demo
5.     namespace: qos-example
6.   spec:
7.     containers:
8.     - name: qos-demo-ctr
9.       image: nginx
10.      resources:
11.        limits:
12.          memory: "200Mi"
13.          cpu: "700m"
14.        requests:
15.          memory: "200Mi"
16.          cpu: "700m"
```

2. Burstable

需要滿足 2 個條件：

- 不是 Guaranteed Pod。
- Pod 內至少有一個容器設置了 memory 或 cpu 的 limits 或 requests。

```
1.   apiVersion: v1
2.   kind: Pod
```

```
3.   metadata:
4.     name: qos-demo-2
5.     namespace: qos-example
6.   spec:
7.     containers:
8.     - name: qos-demo-2-ctr
9.       image: nginx
10.      resources:
11.        limits:
12.          memory: "200Mi"
13.        requests:
14.          memory: "100Mi"
```

3. BestEffort

如果一個 Pods 內的所有容器都沒有設置 request 和 limit，則這個 pod 被認為
是 best-effort。

```
1.   apiVersion: v1
2.   kind: Pod
3.   metadata:
4.     name: qos-demo-3
5.     namespace: qos-example
6.   spec:
7.     containers:
8.     - name: qos-demo-3-ctr
9.       image: nginx
```

查詢以上產生出來的 qosClass：

```
1.   kubectl get pod qos-demo-3 --output=yaml
2.   ---------
3.   spec:
4.     containers:
```

```
5.      ...
6.      resources: {}
7.      ...
8.    status:
9.      qosClass: BestEffort / Burstable / Guaranteed
```

▶ 21.4 Resource 設定的排列組合

從前面的介紹中，我們可以發現 Kubernetes 的 QoS（Quality of Service）主要就是根據 Request / Limit 參數的設定方式來決定。大致上，我們可以將其分為下列四種情況：

	Request	Limit
只設定 Request	O	X
只設定 Limit	X	O
同時設定 Request 和 Limit	O	O
不設定	X	X

1. 只設定 Request：Kubernetes 會保證為其分配至少設定的 request 資源，但不會限制容器的資源使用上限。這個設定通常適用於需要隨時調整資源使用的應用，例如 Web 應用或測試應用，因為它們的資源需求可能會因為用戶流量的變化而波動。

2. 只設定 Limit：在這種情況下，Kubernetes 將會預設使 Request 等於 Limit，從而確保容器的 QoS 等級為 Guaranteed。這意味著 Kubernetes 會限制容器的資源使用不超過設定的 Limit，並且保證分配等於 Limit 的資源。這種設定提供了最大的穩定性，但可能會降低資源的使用效率。

3. 同時設定 Request 和 Limit：在這種設定下，Kubernetes 會為容器保證至少 Request 數量的資源，並限制其資源使用不超過 Limit。這種設定能確保資源的彈性使用，同時防止資源的過度使用。這對於需要彈性資源但又需要防止資源耗盡的應用，例如 Web 應用或 API 服務，非常適合。

4. 不設定：這種情況下，容器可以無限制地使用 CPU 和記憶體資源，這將使得容器的 QoS 等級為 BestEffort。這種設定允許容器使用叢集上任何可用的資源，但可能導致資源分配的不穩定。這種設定通常適用於可以自行管理資源，且不需保證最佳性能的應用，例如測試工具或開發環境。

21.5 實戰心得分享

儘管 Kubernetes 最佳實踐規定應該始終在工作負載上設置資源限制和請求，知道每個應用程序該使用哪些值並不容易。這是一個長期抗戰，而我們可以藉助以下一些實際經驗法則總結一些小建議：

- 盡可能的設定 Request 以及 Limit ，否則在檢視資源利用率時，將會是意義不大的 100%。
- 對不熟悉的應用程式擬定較為通用寬鬆的預設資源分配。
- 在 Namespace 資源隔離的特性下：

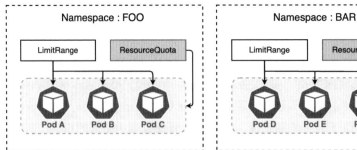

▲ 圖 21-3 namespace

- 使用 ResourceQuota 來限定該 Namespace 的資源使用額度。
- 使用 LimitRange 來宣告在其 Namespace 下的容器資源預設值。
- 使用第三方服務監控或者是資源推薦服務，像是利用 Kubecost、Opencost 來定期檢視資源利用率是否健康。
- 同樣的應用服務，在資源利用率低跟高之間，可能代表著是一兩台的主機的差距，也是成本優化上巨大的關鍵。
- Helm 上的套裝服務也可能預設不符合自身需求的資源分配，像在測試環境中的 ELK 服務並不需要預設般高的資源分配。
- 善用各種工具找出在叢集中無人維護的服務。

參考資料

- 配置 Pod 的服務質量
  https://kubernetes.io/zh-cn/docs/tasks/configure-pod-container/quality-service-pod/

- Kubernetes 資源分配 (limit/request)
  https://developer.aliyun.com/article/679986

- 【譯】Kubernetes 中的資源分配
  https://www.modb.pro/db/46091

從異世界歸來發現只剩自己不會 Kubernetes
初心者進入雲端世界的實戰攻略！

CHAPTER

22

Kubernetes Resources — Namespace

前一章我們藉由 Request / Limit 當作認識 Kubernetes 資源分配的一塊入門磚，本章我們將做一些實戰操作模擬工作中團隊開發的實際情況。首先我們需要認識 Kubernetes 提供的一種叢集中資源劃分並互相隔離的 Group —— Namespaces，並且在 Namespace 底下設置我們的資源分配。

⫸ 22.1 Namespace 是什麼以及何時使用？

Kubernetes 提供了抽象的 Cluster（虛擬叢集）概念，使我們能夠基於各種不同的需求（例如專案的差異性、執行團隊的多元性或商業考量），將原本包含實體資源的單一 Kubernetes Cluster 劃分為多個抽象的虛擬叢集，這就是所謂的 Namespace（命名空間）。

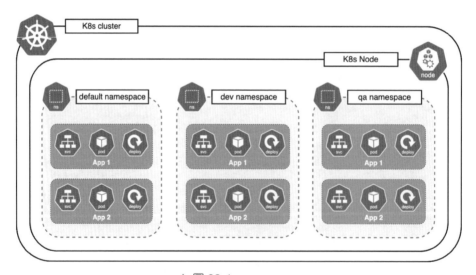

▲ 圖 22-1 namespace

這樣的設計尤其適合需要管理多個跨團隊或跨專案的場景。對於只有少數幾個到十幾個使用者的叢集，可能並不需要建立或使用 Namespace。

NOTE

這就像知名前端框架 Vue 對 Vuex 套件的見解一樣精闢：「它就像一副眼鏡，只有
在你真正需要的時候，你才會想起它」。

22.2 實戰演練

在 Kubernetes 中，命名空間（Namespace）提供一種機制，將同一叢集中
的資源劃分為相互隔離的組別。在同一命名空間內的資源名稱必須是唯一
的，但跨命名空間時則無此要求。命名空間的作用範圍只針對帶有命名空
間的物件（例如 Deployment、Service 等），而對於叢集範圍的物件（例如
StorageClass、Node、PersistentVolume 等）則不適用。接著就來點實際操作
幫助我們更加了解命名空間。

1. 查看 Namespace

```
1.   kubectl get namespace
2.   --------
3.   NAME                    STATUS    AGE
4.   default                 Active    36h
5.   kube-node-lease         Active    36h
6.   kube-public             Active    36h
7.   kube-system             Active    36h
8.   kubernetes-dashboard    Active    36h
```

Kubernetes 會建立四個初始命名空間：

- default：沒有指明使用其它命名空間的物件所使用的預設命名空間。
- kube-system：Kubernetes 系統建立物件所使用的命名空間。

- kube-public：這個命名空間是自動建立的，所有用戶（包括未經過身份驗證的用戶）都可以讀取它。這個命名空間主要用於叢集使用，以防某些資源在整個叢集中應該是可見和可讀的。這個命名空間的命名方式只是一種約定，而不是硬性要求。
- kube-node-lease：該命名空間存放了每個節點相關的 Lease 物件。這些 Lease 物件使得 kubelet 可以發送心跳，這樣 Kubernetes 的控制平面就可以偵測節點的健康狀態。

2. 並非所有資源物件都在 Namespace 中

大多數 Kubernetes 資源（例如 Pod、Service、副本控制器等）都位於某些命名空間中。然而，命名空間資源本身並不在命名空間中。同時，少數的底層資源，例如節點和持久化卷並不屬於任何命名空間。

3. 查看哪些 Kubernetes 資源在命名空間中，哪些不在命名空間中

```
1.    位於命名空間中的資源
2.    kubectl api-resources --namespaced=true
3.
4.    不在命名空間中的資源
5.    kubectl api-resources --namespaced=false
```

4. 建立 Namespace

```
1.    kubectl create namespace demo-namespace
2.    ---------
3.    namespace/demo-namespace creatednamespace
```

接著就可以來查看新建的 demo-namespace：

```
1.    kubectl get namespace
2.    ---------
3.    NAME                    STATUS    AGE
4.    default                 Active    37h
```

```
5.   demo-namespace          Active   7s
6.   kube-node-lease         Active   37h
7.   kube-public             Active   37h
8.   kube-system             Active   37h
9.   kubernetes-dashboard    Active   36h
```

5. 在請求中指定 Namespace

要為當前請求設置命名空間,請使用 --namespace 參數,如此一來回傳的資料將會只有指定 namespace 下的資源,如果沒有指定命名空間則預設為使用 default namesapce。

例如:

```
1.   # 預設使用 default namespace
2.   kubectl get pods
3.   # 使用指定 namespace
4.   kubectl run nginx --image=nginx --namespace=demo-namespace
5.   kubectl get pods --namespace=demo-namespace
```

如果想要指定請求全命名空間的資源,則需要使用 --all-namespaces 參數:

```
1.   kubectl get pods --all-namespaces
```

> 這裡我們可以進階使用 --namespace 縮寫參數 -n 來達到同樣的效果,而 --all-namespace 的縮寫參數則為 -A。

6. 設定預設 Namespace

在我們日常的 kubectl 指令中,使用資源的預設 Namespace 都是 default,如果想要取得其他 Namespace 資源,需要使用 --namespace=<namespace> 參數,我們還有另一個選擇就是修改預設的 Namespace,以用於對應上下文中所有後續 kubectl 指令。

```
1.   kubectl config set-context --current --namespace=demo-namespace
```

就讓我們來驗證一下是否更改成功：

```
1.  kubectl config view --minify | grep namespace:
```

> 只有在預設 Namespace 不為 default 才會出現命名空間欄位。

7. 為 Pod 指定 Namespace

在我們撰寫 Pod 的設定檔中，可以在 metadata.namespace 欄位指定要其運行在哪一個 Namespace 中，如果沒有特別設定將會視預設值而定。

```
1.  apiVersion: v1
2.  kind: Pod
3.  metadata:
4.    namespace: <ns-name>
5.    name: <pod-name>
6.  …
```

22.3 一些 Namespace 的特性

在 Kubernetes 中，命名空間（Namespace）是一個重要的概念，它可以用來將叢集資源分隔為多個獨立的部分。以下將進一步談論命名空間的一些特性：

- 名稱唯一性：在同一個命名空間中，每一種資源的名稱必須是唯一的。例如，兩個服務不能在同一個命名空間中共享同一個名稱。
- 跨命名空間名稱共享：然而，在不同的命名空間中，可以有相同名稱的資源。這意味著你可以在一個命名空間中有一個名為「my-service」的服務，並在另一個命名空間中有另一個同名的服務。
- 刪除命名空間及其資源：當刪除一個命名空間時，該命名空間下的所有資源也將被刪除，這包括該命名空間中的所有 Pod、服務、部署和任何其他關聯的資源。

- 資源配額和限制：Kubernetes 提供了 ResourceQuota 和 LimitRange 兩種
  強大的工具，允許你為每個命名空間分配和限制資源。ResourceQuota 可以
  用來設定每個命名空間的資源使用上限，例如 CPU 時間或記憶體用量。而
  LimitRange 則可以用來限制每個命名空間中 Pod 或容器的資源使用，確保
  單一資源不會佔用過多的共享資源。

筆者碎碎念

命名空間（Namespace）是一個極具價值的特性，它為管理叢集提供了巨大的彈
性。透過劃分命名空間，能夠將同一個叢集的資源進行隔離，使得不同的應用、
部門或專案可以在同一個叢集中獨立運作，而無須擔心資源的混淆或衝突。這一
點在需要管理多個不同的專案時尤其重要。

參考資料

- 為命名空間配置記憶體和 CPU 配額
  https://kubernetes.io/zh-cn/docs/tasks/administer-cluster/manage-resources/quota-memory-cpu-namespace/

- 命名空間配置預設的記憶體請求和限制
  https://sean22492249.medium.com/kubernetes-namespace-%E7%B0%A1%E5%96%AE%E4%BB%8B%E7%B4%B9-c48386949844

- Namespaces
  https://kubernetes.io/docs/concepts/overview/working-with-objects/namespaces/

從異世界歸來發現只剩自己不會 Kubernetes
初心者進入雲端世界的實戰攻略！

CHAPTER

23

Kubernetes Resources — Resource Management

在前兩章，我們已經率先認識了 Request / Limit 和 Namespace 的概念。可以很快地聯想到 ResourceQuota 和 LimitRange 這兩個相關的觀念。其中，ResourceQuota 能夠定義並宣告單一服務的資源使用上限。而 LimitRange 則設定了某一範疇內的資源使用標準和規範。這兩者共同工作，確保資源在 Kubernetes 中的分配既高效又符合規定，避免任何不必要的資源浪費或不當使用。

在預設情況下，Kubernetes 在容器的運行上使用的資源是沒有受到限制的，而身為管理者的我們可以利用 Namespace 作為單位來限制和管理資源的使用，接下來就要繼續深入資源的進階觀念 LimitRange 和 ResourceQuota。

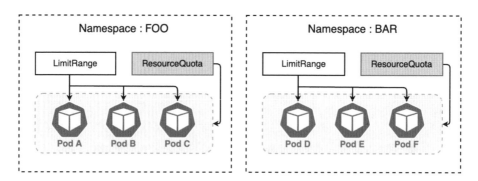

▲ 圖 23-1 LimitRange/ResoruceQuota

▶ 23.1 什麼是 LimitRange？

在 Kubernetes 中，依附在命名空間（Namespace）下的 LimitRange，允許管理員對該命名空間下的單個 Pod 或資源進行資源配額的限制，包括 CPU 或記憶體用量等。這是極為重要的，因為如果一個 Pod 的資源使用沒有被適當地限制，那麼它可能會壟斷整個節點的可用資源。因此，LimitRange 的存在就是為了在 Namespace 內對資源分配實施策略。

具體來説，LimitRange 能實現以下的限制：

- 對每個 Pod 或 Container 在一個 Namespace 中的最小和最大 Request / Limit 進行限制。
- 設定一個 Namespace 中對計算資源的預設 Request / Limit，並且在運行時自動注入到多個 Container 中。
- 對每個 PersistentVolumeClaim 在一個 Namespace 中能申請的最小和最大儲存空間大小進行限制。

23.2 什麼是 ResourceQuota？

在 Kubernetes 中，為了確保多用戶或多團隊在一個叢集中共享資源時能夠均等且公平地使用資源，ResourceQuota 來到了我們的視野。透過定義一個 ResourceQuota 物件，管理員可以限制每個命名空間內的總資源消耗量。此外，它還能限制在命名空間內可以建立的物件數量。同時，也能控制該命名空間中的資源可以消耗的計算資源總量。

具體來説，一個 ResourceQuota 能實現以下的策略：

- 管理員為每個命名空間建立一個 ResourceQuota。
- 用戶在命名空間中建立資源（例如 pods、services 等），配額系統追蹤使用情況，以確保它不超過 ResourceQuota 中定義的資源限制。
- 如果建立或更新資源違反了配額限制，請求將會收到 HTTP 403 FORBIDDEN 狀態碼的失敗回應，並附帶解釋違反的限制的消息。

ResourceQuota 在 Kubernetes 中提供了不止對運算資源的限額功能。

以下列舉了它可以控制的不同類型的資源：

- 運算資源配額（Compute Resource Quota）：
  - limits.cpu：所有非終端狀態的 pods 的 CPU 限制總和。
  - limits.memory：所有非終端狀態的 pods 的記憶體限制總和。
  - requests.cpu：所有非終端狀態的 pods 的 CPU 請求總和。
  - requests.memory：所有非終端狀態的 pods 的記憶體請求總和。
  - hugepages-<size>：所有非終端狀態的 pods 的特定大小的大頁請求數不能超過此值。

- 擴展資源配額（Resource Quota For Extended Resources）
  - 擴展資源不允許超預設值，所以只允許使用 requests. 前綴的配額項目。例如，對於 GPU 資源（資源名稱為 nvidia.com/gpu），如果你希望限制命名空間中請求的 GPU 總數為 4，你可以這樣定義配額：requests.nvidia.com/gpu: 4

- 儲存資源配額（Storage Resource Quota）
  - 除了可以限制特定命名空間中可以請求的儲存資源總和外，還可以基於關聯的儲存類型來限制儲存資源的消耗。例如：
    - requests.storage：所有持久性卷請求的儲存請求總和。
    - <storage-class-name>.storageclass.storage.k8s.io/requests.storage：所有特定儲存類型持久性卷請求的請求總和。

- 物件計數配額（Object Count Quota）
  - 可以對所有標準的、有命名空間的資源類型設定配額，以限制某個命名空間中的資源總數。例如，你可能希望限制一些 Secret 數量。對於這種類型的配額，存在於伺服器儲存中的物件將被計入配額。

⧉ 23.3 實戰演練 — LimitRange

現在我們已經對 Kubernetes 的 LimitRange / ResourceQuota 有了基本的認識。接下來，讓我們進一步去探索 LimitRange / ResourceQuota 的實際應用，並了解當資源使用達到其上限時會發生什麼。

首先，我們將建立一個 LimitRange 的設定，並將其應用於我們新建的 Namespace。在這個設定中，我們將為 CPU 和記憶體設定一個最大使用量。接著，我們將在此 Namespace 中建立一個 Pod，並觀察其資源使用情況。

然後，我們將嘗試讓這個 Pod 的資源使用超過我們設定的上限，看看會發生什麼事吧！

1. 建立 Namespace

```
1.   kubectl create ns demo-namespace
2.   -------
3.   namespace/demo-namespace created
```

將先建立好的 Namespace 設定為預設值：

```
1.   kubectl config set-context --current --namespace=demo-namespace
2.   --------
3.   Context "docker-desktop" modified
```

2. 建立宣告預設資源配額的 LimitRange

範例檔 **limit-range.yaml**

```
1.   # limit-range.yaml
2.   apiVersion: v1
3.   kind: LimitRange
4.   metadata:
```

```
5.    name: limit-range
6.  spec:
7.    limits:
8.     - max:
9.         cpu: 1000m
10.        memory: 500Mi
11.      min:
12.        cpu: 100m
13.        memory: 50Mi
14.      type: Container
```

建立 LimitRange：

```
1.  kubectl apply -f limit-range.yaml
2.  -------
3.  limitrange/limit-range created
```

查看一下剛剛建立的 LimitRange：

```
1.  kubectl get limitrange limit-range --output=yaml
2.  -------
3.  .....
4.  limits:
5.  - default:
6.      cpu: "1"
7.      memory: 500Mi
8.    defaultRequest:
9.      cpu: "1"
10.     memory: 500Mi
11.    max:
12.     cpu: "1"
13.     memory: 500Mi
14.    min:
15.     cpu: 100m
```

```
16.       memory: 50Mi
17.    type: Container
```

建立了 LimitRange 後，在此 Namespace 下建立 Pod 時，如果沒有特別宣告自己的資源請求設定，Kubernetes 就會依照 LimitRange 設定提供預設的請求和限制，如果該資源已有設定資源配額，則 Kubernetes 會阻止建立超出這些規範的資源。

- 如果 Pod 內的任何容器沒有宣告自己的請求以及限制，即為該容器設置預設的 CPU 和記憶體請求或限制。
- 確保每個 Pod 中的容器宣告的請求至少大於等於 limits.defaultRequest。
- 確保每個 Pod 中的容器宣告的請求至少小於等於 limits.default。

3. 建立一個沒有宣告請求和限制的 Pod

```
1.   # limit-range-pod.yaml
2.   apiVersion: v1
3.   kind: Pod
4.   metadata:
5.     name: limit-range-pod
6.   spec:
7.     containers:
8.     - name: default-limit-range-pod
9.       image: nginx
```

建立起資源：

```
1.   kubectl apply -f ./limit-range-pod.yaml
2.   -------
3.   pod/limit-range-pod created
```

查看 LimitRange 是否替我們設定了請求與限制：

```
1.   kubectl get pod limit-range-pod --output=yaml
```

```
2.  --------
3.  ....
4.  containers:
5.    - image: nginx
6.      imagePullPolicy: Always
7.      name: limit-range-pod
8.      resources:
9.        limits:
10.          cpu: "1"
11.          memory: 500Mi
12.        requests:
13.          cpu: "1"
14.          memory: 500Mi
15. ....
```

成功看到相關設定。

4. 建立一個超過最大限制或不滿足最小請求的 Pod

當我們嘗試著建立一個超過 LimitRange CPU limit 的資源時，Kubernetes 將會直接回傳以下類似錯誤訊息，因為其中定義了過高的 CPU limit。

```
1.  Error from server (Forbidden): error when creating "examples/admin/
    resource/limit-range-pod.yaml":
2.  pods "limit-range-pod" is forbidden: maximum cpu usage per
    Container is 800m, but limit is 1500m.
```

反之當我們在嘗試著建立一個不滿足 LimitRange CPU request 的資源時，也會看到建立失敗的錯誤訊息：

```
1.  Error from server (Forbidden): error when creating "examples/admin/
    resource/limit-range-pod.yaml":
2.  pods "limit-range-pod" is forbidden: minimum cpu usage per
    Container is 200m, but request is 100m.
```

⫸ 23.4 實戰演練 — ResourceQuota

現在我們進一步探索 ResourceQuota 的實際應用，並看看當超出配額時將會發生什麼事。

首先，我們將建立一個 ResourceQuota，並將其應用到我們新建的 Namespace 中。在這個 ResourceQuota 中，我們將為 CPU 和記憶體設定一個上限，然後我們將在此 Namespace 中建立 Pod 並觀察其行為。

最後，我們模擬出 Pod 的資源使用超過設定配額的情境，看看會發生什麼。

1. 建立 Namespace

```
1.   kubectl create ns quota-namespace
```

2. 將先建立好的 Namespace 設定為預設值

```
1.   kubectl config set-context --current --namespace=quota-namespace
```

3. 建立對運算資源限額的 ResourceQuota

範例檔 **resource-quota.yaml**

```
1.   # resource-quota.yaml
2.   apiVersion: v1
3.   kind: ResourceQuota
4.   metadata:
5.     name: resource-quota
6.   spec:
7.     hard:
8.       requests.cpu: "1"
9.       requests.memory: 1Gi
10.      limits.cpu: "2"
11.      limits.memory: 2Gi
```

建立 ResourceQuota：

```
1.   kubectl apply -f resource-quota.yaml
2.   -------
3.   resourcequota/resource-quota created
```

查看剛建立的 ResourceQuota：

```
1.   kubectl describe resourcequota
2.   -------
3.   Name:              resource-quota
4.   Namespace:         quota-namespace
5.   Resource         Used   Hard
6.   --------         ----   ----
7.   limits.cpu         0      2
8.   limits.memory      0      2Gi
9.   requests.cpu       0      1
10.  requests.memory    0      1Gi
```

4. 建立一個在 ResourceQuota 限額內的 Pod

範例檔 **resource-quota-pod.yaml**

```
1.   # resource-quota-pod.yaml
2.   apiVersion: v1
3.   kind: Pod
4.   metadata:
5.     name: resource-quota-pod
6.   spec:
7.     containers:
8.       - name: resource-quota-pod
9.         image: nginx
10.        resources:
11.          limits:
```

```
12.          memory: "800Mi"
13.          cpu: "800m"
14.        requests:
15.          memory: "600Mi"
16.          cpu: "400m"
```

建立起資源：

```
1.  kubectl apply -f resource-quota-pod.yaml
2.  -------
3.  pod/resource-quota-poddemo created
```

這時再次查看當前 ResourceQuota：

```
1.  kubectl describe resourcequota
2.  -------
3.  Name:              resource-quota
4.  Namespace:         quota-namespace
5.  Resource          Used    Hard
6.  --------          ----    ----
7.  limits.cpu        800m    2
8.  limits.memory     800Mi   2Gi
9.  requests.cpu      400m    1
10. requests.memory   600Mi   1Gi
```

可以發現 Used 的資源已經隨著我們新建的資源增加。

5. 建立一個超過最大限額的 Pod

建立了 ResourceQuota 之後，如果我們在該 Namespace 中建立 Pod 或 PersistentVolumeClaims，並嘗試超出限制，我們將會看到建立失敗的錯誤訊息。

例如，如果我們嘗試建立一個超過 CPU、Memory 限額的 Pod，或者試圖建立超過限制的 PersistentVolumeClaims，將會看到類似以下的錯誤：

```
1.  Error from server (Forbidden): error when creating "examples/
    admin/resource/quota-mem-cpu-pod-2.yaml": pods "quota-mem-cpu-
    demo-2" is forbidden: exceeded quota: mem-cpu-demo, requested:
    requests.memory=700Mi,used: requests.memory=600Mi, limited:
    requests.memory=1Gi
```

這樣，我們就可以清楚地看到 ResourceQuota 的影響，並了解到在一個具有配額的 Namespace 中，我們無法超出限制來建立資源。

筆者碎碎念

Namespace 提供了一種便利的方式來進行資源分配。當我們結合使用 LimitRange 和 ResourceQuota 等基於命名空間的資源分配時，能為不同的部門靈活規劃相對應的資源分配。

ResourceQuota 的使用方法與 Limit / Request 的概念相似，它是用來控制 Namespace 資源使用量的上限，包括 CPU、記憶體等。一旦理解了這個基本觀念，對於各種資源分配的理解都會如行雲流水。

參考資料

- 配置 Pod 的服務質量：
  https://kubernetes.io/zh-cn/docs/tasks/configure-pod-container/quality-service-pod/

- Kubernetes 資源分配 (limit/request)：
  https://developer.aliyun.com/article/679986

- 【譯】Kubernetes 中的資源分配：
  https://www.modb.pro/db/46091

Kubernetes Resources —
Metrics Server

到目前為止，我們學習了許多關於資源設定的觀念，但漸漸地會發現一件事：
我們該如何知道以及監控所有服務的資源利用率和健康狀況等等。Kubernetes
有許多指標資料需要收集，大致上可以分為叢集本身以及 Pod，包含節點是否
正常運行，像是 Disk、CPU、Memory 利用率和需要跟 Kubernetes 獲取的部署
副本數、存活監測、健康監測，這些都需要一個工具來替我們收集資源指標並
且整合，而 Kubernetes 本身沒有提供這樣的工具，需要使用擴充套件來實現。

▶ 24.1 Metrics Server 是什麼？

Metrics Server 在 Kubernetes 中扮演著關鍵的角色，尤其在監控資源使用情況
和自動伸縮工作負載時。儘管 Docker Desktop 提供的 Kubernetes 沒有預設安
裝 Metrics Server，但在像是 Google Kubernetes Engine（GKE）的環境中，
Google 早已提供了資源監控服務。

Metrics Server 是一種 Kubernetes 的叢集級別資料聚合器（Aggregator）。它主
要是透過將 kube-aggregator 部署到 API Server 上，並基於 kubelet 從各節點
收集資源使用的指標資料，然後再將這些資料儲存在 Metrics Server 的記憶體
中。值得注意的是，因為資料是儲存在記憶體中，因此 Metrics Server 並不會
保留歷史資料。換句話說，每當 Metrics Server 重啟時，之前的資料就會消失。

Metrics Server 提供的這些資源使用指標，使我們可以透過 Kubernetes API
Server 的 /apis/metrics.k8s.io 端點進行訪問。這種設計使得 Metrics Server 成
為了 Kubernetes 叢集資源監控與管理的重要工具。

Metrics Server 不只是用來收集資源利用資料，它也是實現 Kubernetes 叢集中
Horizontal Pod Autoscaler（HPA）和 Kubernetes Autoscaling 的基礎服務。
HPA 使用 Metrics Server 來決定是否需要自動擴展或縮小工作負載，以滿足變
化的需求。

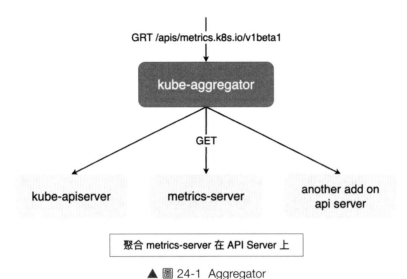

▲ 圖 24-1 Aggregator

此外，Metrics Server 是 Kubelet 中 resource metrics API 和 custom metrics API 的一部分。resource metrics API 提供了 Pod 和 Node 級別的 CPU 和記憶體使用指標，這對於 Kubernetes 的 Scheduling decisions 和 Kubelet 的 evictions 是至關重要的。

相較於其他的監控解決方案如 Prometheus，Metrics Server 的焦點在於即時（Real-Time）或近即時的資源利用率監控，並且為了簡化與效能，不提供長期儲存或資料查詢語言等功能。這使得 Metrics Server 更適合為 Kubernetes 的內建自動擴展系統提供資料。

最後，儘管 Metrics Server 可以運行在任何 Kubernetes 叢集上，但要注意的是，對於一些特定的環境（例如 AWS 或 Google Cloud），可能需要進行特定的設定才能讓 Metrics Server 正確運作。例如，你可能需要為 Metrics Server 提供特定的權限，或者在安裝時設定特定的 flags。

24.2 Metrics Server 原理

Metrics Server 會定期地向 Kubelet 的 Summary API（例如 /api/v1/nodes/node_name/stats/summary）抓取指標資料，然後將這些資料進行整合後儲存在內部記憶體中，並透過 metric-api 來提供給外部存取。

為了能夠實現這樣的功能，Metrics Server 利用了 api-server 的部分函式庫，例如認證（Authentication）和 API 版本控制等等。不過，為了能在記憶體中儲存這些資料，Metrics Server 沒有採用預設的 etcd 儲存方式，而是採用了基於 Storage interface 的記憶體儲存。

> **NOTE**
>
> 需要注意的是，由於所有資料都存放在記憶體中，這些監控資料是不具備持久性的。但你可以選擇透過第三方儲存服務來實現資料的持久化。

來看看 Metrics-Server 的架構如何運作：

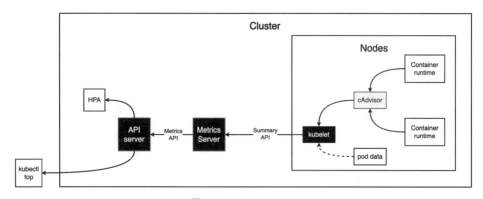

▲ 圖 24-2 Metrics Server

Metrics Server 會從 Kubelet、cAdvisor 等進行資料收集，然後提供給像是 Dashboard、HPA 控制器等服務使用。基本上，Metrics Server 的角色就是進行資料轉換，它將 cAdvisor 格式的資料轉換成了 Kubernetes API 可讀的 JSON 格式。所以我們可以推斷，Metrics Server 的程式碼中應該實作了從 Metric 取得所有接口訊息並解析出資料的步驟。因此當我們發送請求給 Metrics-Server 時，它已經從 cAdvisor 定期取得了資料，然後就會直接回傳這些存在快取中的資料。而 Metrics Server 則會透過 kubelet 來獲得從 cAdvisor 捕捉的資料。

NOTE

在 Kubernetes 1.7 版本之前，Kubernetes 在每個節點都安裝了一個名為 cAdvisor 的程式，該程式負責獲取節點和容器的 CPU、記憶體等資料。然而，從 1.7 版本開始，Kubernetes 將 cAdvisor 精簡並內建於 kubelet 中，因此現在我們可以直接從 kubelet 中獲取這些資料。

24.3 安裝 Metrics Server

現在我們可以使用官方提供的最新版本載點安裝：

```
1.  kubectl apply -f https://github.com/kubernetes-sigs/metrics-server/
    releases/latest/download/components.yaml
```

可能有些朋友會直接照著官方安裝指示下載但卻失敗，其中一個原因可能是沒有開啟連結中設定檔第 136 行左右的設定 --kubelet-insecure-tls ，此參數可以讓 Metrics Server 停用憑證驗證，畢竟我們在本地開發環境中練習，不需要額外使用憑證。

大致的內容如下：

```
1.  apiVersion: apps/v1
2.  kind: Deployment
3.  metadata:
4.    labels:
5.      k8s-app: metrics-server
6.    name: metrics-server
7.    namespace: kube-system
8.  ......
9.      spec:
10.       containers:
11.       - args:
12.           - --cert-dir=/tmp
13.           - --secure-port=4443
14.           - --kubelet-preferred-address-types=InternalIP,
    ExternalIP,Hostname
15.           - --kubelet-use-node-status-port
16. =================
17.           # 加上這個  Do not verify the CA of serving certificates
    presented by Kubelets. For testing purposes only.
18.           - --kubelet-insecure-tls
19. =================
20.         image: k8s.gcr.io/metrics-server/metrics-server: v0.4.2
21.         imagePullPolicy: IfNotPresent
```

也可以直接使用這邊修改好的設定檔：

```
1.  kubectl apply -f https://raw.githubusercontent.com/MikeHsu0618/
    kubernetes-from-another-world/main/ch24/components.yaml
2.  -------
3.  serviceaccount/metrics-server created
4.  clusterrole.rbac.authorization.k8s.io/system: aggregated-metrics-
```

```
     reader created
 5.  clusterrole.rbac.authorization.k8s.io/system: metrics-server created
 6.  rolebinding.rbac.authorization.k8s.io/metrics-server-auth-reader
     created
 7.  clusterrolebinding.rbac.authorization.k8s.io/metrics-server:
     system: auth-delegator created
 8.  clusterrolebinding.rbac.authorization.k8s.io/system: metrics-server
     created
 9.  service/metrics-server created
10.  deployment.apps/metrics-server created
11.  apiservice.apiregistration.k8s.io/v1beta1.metrics.k8s.io created
```

成功在 kube-system 中建立：

```
1.  kubectl get pods -n kube-system | grep metrics-server
2.  --------
3.  metrics-server-9f897d54b-l2rc4       1/1     Running    0      5m11s
```

24.4 顯示資源使用訊息

kubectl top 可以查看節點和 Pod 的資源使用訊息，包含 node 和 pod 兩個子指令，可以顯示相關的資源佔用率。

```
1.  kubectl top pods -n kube-system
2.  -------
3.  NAME                                        CPU(cores)    MEMORY(bytes)
4.  coredns-6d4b75cb6d-pp56z                    6m            0Mi
5.  coredns-6d4b75cb6d-qk4vm                    6m            0Mi
6.  etcd-docker-desktop                         28m           0Mi
7.  kube-apiserver-docker-desktop               59m           0Mi
8.  kube-controller-manager-docker-desktop      32m           0Mi
```

```
9.   kube-proxy-8j6xq                          1m        0Mi
10.  kube-scheduler-docker-desktop             6m        0Mi
11.  metrics-server-9f897d54b-l2rc4            8m        0Mi
12.  storage-provisioner                       4m        0Mi
13.  vpnkit-controller                         1m        0Mi
```

筆者碎碎念

藉由擁有這類資源指標收集工具，我們已經在資源監控領域取得重要的突破。透過監控，我們不僅可以對整個叢集的狀態有更深的了解，更可以依據不同的資源指標設定多種自動化策略。其中，我們即將討論的「自動伸縮」（AutoScaling）功能就是一個典型的例子。然而，在開始使用自動伸縮功能之前，我們必須先安裝和設定好 Metrics Server，這是開始使用任何自動伸縮策略的基礎要求。透過深化了解並善用 Metrics Server，我們將能夠更有效地實現 Kubernetes 叢集的自動化管理。

參考資料

- Kubernetes Metrics Server
  https://kubernetes-sigs.github.io/metrics-server/

- Kubernetes 核心指標監控——Metrics Server
  https://www.cnblogs.com/zhangmingcheng/p/15770672.html

- 資源指標 metrics-server
  https://ithelp.ithome.com.tw/articles/10241138

Part 8
主題篇—AutoScaling

身為 Server 守護者的
你是不是也沒辦法
睡個好覺？

面對低落的資源利用率萌生縮小規格的念頭時，
腦中就會浮現出曾經突然暴衝的工作負載而不敢
輕舉妄動。

很明顯的，Kubernetes 提供我們一個輕量且彈性
的解決方案。

Kubernetes AutoScaling — AutoScaling 是什麼？

目前我們已深入探討資源設定和監控的各種知識。但僅僅瞭解這些資源指標，並且還要手動去調整，其實並沒有完全發揮它們的價值。為了更好地利用這些資源指標，「自動化資源調整」（或稱為 AutoScaling）應運而生。簡而言之，AutoScaling 會根據你的資源設定並實時監控資源使用狀況，依照系統的負載變化自動地進行水平擴展、垂直擴展或多維擴展，甚至是縮減。但要特別注意的是，所有這些自動操作都建立在已正確設定的資源設定和 Metrics Server 上。因此，要想真正實現自動化管理，深入了解並妥善運用 Metrics Server 和資源設定是關鍵。

25.1 Autoscaler 的種類

在 Kubernetes 中，自動伸縮的能力是由幾種不同的 Autoscaler 提供的，這些 Autoscaler 每一種都有其獨特的用途和功能。

以下是我們接下來將會深入探討的幾種 Autoscaler：

1. Cluster Autoscaler（CA）：Cluster Autoscaler 負責根據叢集負載的增減，自動調整節點數量。

2. Horizontal Pod Autoscaler（HPA）：HPA 是 Kubernetes 中的一個自動化元件，負責根據指定的指標，例如 CPU 和 Memory 使用率，動態調整 Pod 的實例數量。HPA 的主要目的是在負載增加時自動擴展 Pod 的數量，並在負載減少時縮減，確保資源的高效利用並維持應用的響應能力。

3. Vertical Pod Autoscaler（VPA）：與 HPA 不同，VPA 會根據負載情況自動調整 Pod 的 CPU 和記憶體請求。這意味著它可以自動為 Pod「縮小或放大」資源設定，以最佳化其效能。

4. Multidimensional Pod Autoscaler（MPA）：Multidimensional Pod Autoscaler 是另一種自動伸縮方法，其結合了 HPA 和 VPA 的優點，能夠基

於 CPU、記憶體等多維度資源的使用情況進行自動伸縮。MPA 可以同時進行水平和垂直的自動伸縮，對於需要更全面和靈活的自動調整方案，MPA 會是一個好選擇。

5. Custom Pod Autoscaler（CPA）：對於一些特殊的自動縮放需求，例如需要根據業務指標（例如訂單數量、佇列長度等）來進行擴縮的情況，我們可能需要使用自訂的 Pod Autoscaler。這時，一些工具如 Kubernetes Event-driven Autoscaling（KEDA）或 Prometheus adapter 就顯得格外重要。KEDA 可以依據事件驅動的方式來進行自動擴展，非常適合處理異步的工作負載；而 Prometheus Adapter 則可以讓我們根據 Prometheus Metrics 來進行擴展或縮減。

25.2 Cluster Autoscaler（CA）

Cluster Autoscaler 的主要工作為調整 node-pool 的大小。作為一個叢集級別的 autoscaler，它能在系統負載增加時添加新的節點，並在負載減少時關閉不必要的節點。

- Scale-up：當有 Pod 的狀態為 unschedulable 時，Cluster Autoscaler 將在大約十秒後迅速判斷是否需要進行水平擴展。請注意，雖然判斷動作可能在十秒內完成，但實際的機器啟動時間可能需要數分鐘到十多分鐘才能處於可用狀態。
- Scale-down：Cluster Autoscaler 會定期（預設每十秒）檢查 CPU 及記憶體請求的總和是否低於預設值，同時確保沒有任何 Pod 或 Node 的調度限制。

有些設定如果不妥當，可能會導致 CA 無法進行自動縮減。例如，當 CA 試圖關閉節點並遷移 Pod 時，可能會違反 Pod 的親和性／反親和性規則或 PodDisruptionBudget。 在 節 點 上 設 定 annotation: "cluster-autoscaler.kubernetes.io/scale-down-disabled": "true"，可以防止節點被縮減。

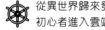
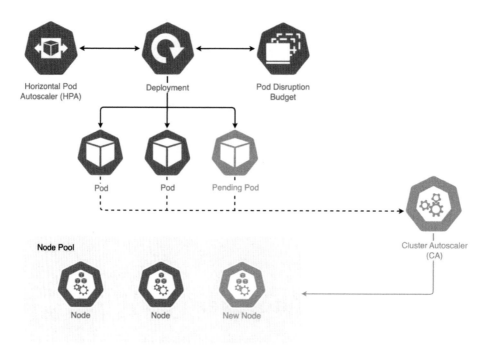

▲ 圖 25-1　cluster autoscaler

正如前文所述，Cluster Autoscaler 是叢集級別的調節器，因此在本地只有一個
節點的 Docker Desktop 環境中可能無法體驗到它的強大。如果我們使用的是雲
端平台級別的 Kubernetes，例如 Google Cloud Platform 的 GKE 或 Amazon 的
EKS，這些平台能提供更全面的資源，並進行更高級別的自動調度。

25.3 Horizontal Pod Autoscaler（HPA）

Horizontal Pod Autoscaler（HPA）屬於 Pod 層級的自動調節工具（autoscaler），
主要負責調整 Pod 的數量，以應對系統負載的波動。具體來說，當負載增加
時，autoscaler 會自動擴展 Pod 的數量直到達到設定的最大值；反之，當負載
減少時，autoscaler 則會自動縮減 Pod 的數量直到達到設定的最小值。

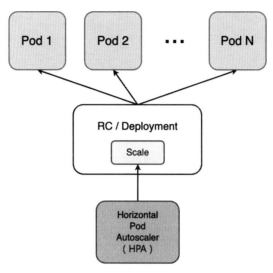

▲ 圖 25-2 horizonal pod autoscaler

- Scale-up：autoscaler 會定期檢查 Metrics Server 的資料。若發現系統的使用率超過了預設的閾值，autoscaler 就會透過增加 replicas 數值，來擴展 Pod 的數量。

- Scale-down：同樣地，autoscaler 也會定期檢查 Metrics Server 的資料，並根據使用率低於預設閾值來減少 replicas 數值，進行縮減。

如果在 Deployment 中已經設定了 replicas 數值，那麼這個設定將會被 HPA 的 replicas 設定所覆蓋。換句話說，HPA 會優先根據自身的設定來調整 Pod 的數量，而不會考慮原本資源物件的 replicas 設定。

當系統完成擴展或縮減後，autoscaler 會等待約三到五分鐘讓系統穩定下來，然後再開始檢查 Metrics Server 的資料。

使用率的簡略計算方式如下：假設目前的 metric 值（currentMetricValue）為 200m，而設定的 metric 值（desiredMetricValue）為 100m，則代表目前需要將 replicas 的數量增加 200/100 = 2 倍。

TIPS

具體的計算公式為：

desiredReplicas = ceil[currentReplicas * (currentMetricValue / desiredMetricValue)]

除此之外，我們也可以設定自定義或外部的 metrics 來觸發自動調節。

 NOTE

需要注意的是，只有在 v2beta2 或更新的 HPA 版本中，才能檢查記憶體的使用情況。在 v1 版本中，HPA 只能檢查 CPU 的使用情況。由於 HPA 的 API 更新得非常快，因此在網路上可以找到許多不同版本（例如 v1、v2、v2beta2 等）的範例教學。但建議大家還是優先參考最新的 API 檔案來了解最新的功能和使用方式。

25.4 Vertical Pod Autoscaler（VPA）

VPA 是一個極其實用的工具，它能自動化地確定並調整 Pod 的資源需求（例如 CPU 和記憶體）。這意味著我們不再需要手動進行監控和設定，這對於缺乏系統優化經驗的 DevOps 新手來說是一大好處。

- Scale-up：VPA 會檢查 metrics，如果發現使用率超過了設定值，則會增加 Pod 的 resources.requests，然後透過重啟 Pod 來實際更新設定。
- Scale-down：與 Scale-up 相反，如果使用率低於設定值，則 VPA 會降低 Pod 的 resources.requests，並透過重啟 Pod 來更新設定。

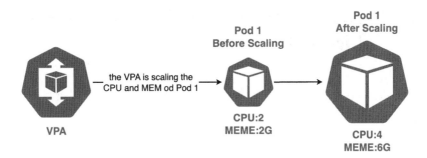

▲ 圖 25-3 vertical pod autoscaler

當 VPA 嘗試在最佳時機進行 Pod 的重啟時，必須刪除原有的 Pod 並建立一個新的已更新 resources.requests/limits 的 Pod。VPA 會參考 Pod 的歷史資料來做出決策。

目前 VPA 和 HPA 還不能混用，除非 HPA 使用 custom metric 作為觸發條件，或是使用了如 Multidimensional Pod Autoscaler（MPA）等工具來實現 VPA 和 HPA 的同時使用。

值得注意的是，VPA 並未被 Kubernetes 列入原生支援的元件，但某些雲端服務提供商已將其納入其服務。例如，Google Kubernetes Engine（GKE）就原生支援 VPA，但 AWS 的 EKS 雖然並未原生支援 VPA，使用者仍可以手動進行安裝。

VPA 就像 Metrics Server 一樣，是一種自定義資源（Custom Resource），這意味著它可能不會被預設安裝在 Kubernetes 中。但是，許多核心功能都依賴這些自定義資源來實現，這讓 Kubernetes 的模組化設計更具彈性和擴展性。

▶ 25.5 Multidimensional Pod Autoscaler（MPA）

Multidim Pod Autoscaler（MPA）是一種特殊的擴縮概念，可以讓我們同時使用多種方式來擴縮我們的叢集。

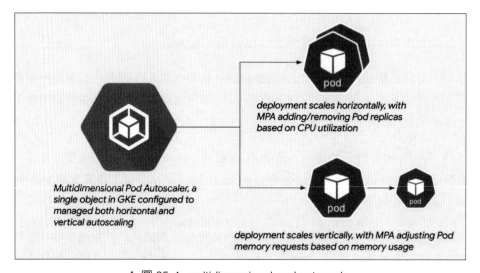

deployment scales horizontally, with MPA adding/removing Pod replicas based on CPU utilization

Multidimensional Pod Autoscaler, a single object in GKE configured to managed both horizontal and vertical autoscaling

deployment scales vertically, with MPA adjusting Pod memory requests based on memory usage

▲ 圖 25-4 multidimensional pod autoscaler

目前只有 Google Cloud Platform（GCP）的 Google Kubernetes Engine（GKE）提供這種多維度的擴縮操作。在欣賞 Google 的創新之餘，我們也可以看出雲端平台與開源社群之間的關係：由於 Kubernetes 開源社群的 autoscaler 專案尚未實現多維擴縮，也許是各大雲端平台都在等待開源社群的進展，然後將新的功能囊括進來。

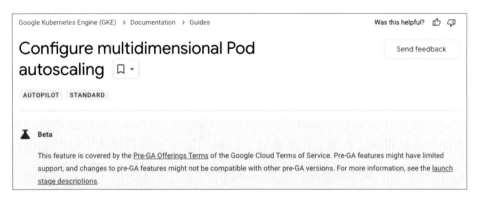

▲ 圖 25-5　multidimensional pod autoscaler-beta

此功能目前僅在 GCP GKE 中提供，並且處於 Beta 版本，所以無法在本地或其他平台上實現。如果你在 GCP GKE 中有非正式環境，並對這項功能有興趣，可以嘗試一下。

多維度擴縮目前只能「根據 CPU 進行 HPA，以及根據記憶體進行 VPA」。因此，如果你在使用 MPA，需要注意在 Deployment 的 resources 中的 cpu requests/limits 是必須要事先設定的欄位，因為目前的 VPA 只能依據記憶體進行調整。

 NOTE

如果你對這個主題有更深的興趣，建議可以查閱官方文件以獲得更多的細節和理解：https://cloud.google.com/kubernetes-engine/docs/how-to/multidimensional-pod-autoscaling。

▶ 25.6 Custom Pod Autoscaler

當我們需要更靈活、更專門的自動縮放解決方案時，Custom Pod Autoscaler（CPA）的價值就體現出來了。不同於原生的 Horizontal Pod Autoscaler（HPA）主要基於 CPU 和記憶體使用率來調整 Pod 的數量，CPA 允許我們根據各種業務指標（例如，訂單數量、佇列長度等）進行擴縮，甚至可以與 Prometheus 這種監控系統整合，根據自訂義指標來調整 Pod 的數量。

例如：

1. Kubernetes Event-driven Autoscaling（KEDA）：KEDA 是一個 CPA 解決方案，它擴展了原生 HPA 的能力，可以基於各種事件來進行自動縮放。KEDA 可以與許多事件溯源整合，例如 Kafka、RabbitMQ，或者雲端提供的消息佇列服務等。這使得 KEDA 能夠在無任何工作任務的情況下將 Pod 縮小到零，並在需要時根據事件的發生而彈性擴展。

2. Prometheus Adapter：這是另一個 CPA 解決方案，它允許我們使用 Prometheus 查詢結果作為 Pod 自動縮放的指標。這種自定義的彈性使得我們可以根據實際的業務需求或性能指標來進行縮放，而不僅僅是基於 CPU 和記憶體使用率。這對於需要根據自定義或特殊指標進行自動縮放的工作負載非常有用。

因此使用 CPA，我們可以滿足更專門、更精細的自動縮放需求，以確保我們的應用程式在面對變化的工作負載時，始終保持最佳狀態，在後續的章節中我們也會有一些實戰演練。

筆者碎碎念

經過這次的探討，我們對於 Kubernetes 的 AutoScaling 方案有了更加深入的認識。從 Horizontal Pod Autoscaler、Vertical Pod Autoscaler、到 Multidimensional Pod Autoscaler 和 Custom Pod Autoscaler，每種 Autoscaler 都各有其獨特的適用場景與調整策略，它們不僅提供了系統資源監控與調配的能力，同時也為我們帶來了大量的靈活性和可能性。

值得注意的是，許多 Kubernetes 的進階功能，例如 Vertical Pod Autoscaler 和 Metrics Server，都屬於自定義資源（Custom Resources）。這些自定義資源並不會預設安裝在 Kubernetes 當中，但它們的存在卻大大擴充了 Kubernetes 的功能範疇，並使其成為一個高度模組化的平台。

隨著 Kubernetes API 的快速迭代，保持對於這些新功能的更新並熟悉其使用方法將是一大挑戰。就像那句開源社群經典名言：「別再更新了，學不動了！」然而，這也正是 Kubernetes 魅力所在。

參考資料

- Configuring multidimensional Pod autoscaling
  https://cloud.google.com/kubernetes-engine/docs/how-to/
  multidimensional-pod-autoscaling

- Scale container resource requests and limits
  https://cloud.google.com/kubernetes-engine/docs/how-to/vertical-pod-
  autoscaling

- Kubernetes 那些事 — Auto Scaling
  https://medium.com/andy-blog/kubernetes-%E9%82%A3%E4%BA
  %9B%E4%BA%8B-auto-scaling-7b887f61fdec

- Kubernetes/autoscaler
  https://github.com/kubernetes/autoscaler

- Kubernetes Horizontal Scaling/Vertical Scaling 概念
  https://sean22492249.medium.com/kubernetes-horizontal-scaling-
  vertical-scaling-%E6%A6%82%E5%BF%B5-e8e70ce6f034

- Kubernetes Autoscaling 相關知識小整理
  https://weihanglo.tw/posts/2020/k8s-autoscaling/

CHAPTER

26

Kubernetes AutoScaling — Horizontal Pod AutoScaler

在前面一章的種類介紹中，我們了解到 Horizontal Pod Autoscaler（HPA）的優越之處就在於其能夠根據實際需求進行自動擴縮，從而有效避免資源閒置或是過度負載的情況發生。常見的使用情境可能包括搶票系統、訂餐系統等，這些服務都會在某個特定的時段面臨大量的服務需求，例如午餐時間的熊貓外送服務的使用者數量就會突增。雖然大部分的尖峰需求時刻是可以預測的，但這並不代表我們的 DevOps 團隊必須每天在特定時間手動調整服務的負載能力。

畢竟，未能預見的尖峰需求時刻總會出現，這時如果沒有及時的應對措施，那就只能無可奈何地看著服務過載、效能下降。幸好我們有 Kubernetes 的 autoscaler，在這種情況下，它能在短短幾十秒內即時將服務擴展，以滿足突增的需求。下面，我們將實際利用 HPA 進行操作練習，以進一步體驗並理解其運作原理與優勢。

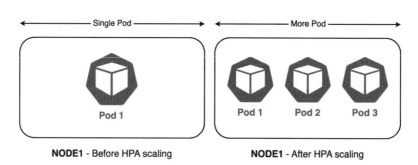

▲ 圖 26-1 HPA

26.1 確認 Metrics Server 是否就緒

在進行之前，我們需要擁有一個已設定 Metrics Server 的 Kubernetes 叢集用來收集各種資源指標當作 autoscaling 的依據。

```
1.  kubectl top node
2.  -------
```

```
3.   NAME              CPU(cores)    CPU%    MEMORY(bytes)    MEMORY%
4.   docker-desktop    258m          6%      5717Mi           72%
```

如果還沒有安裝的朋友可以參考第 24.3 節的〈安裝 Metrics Server〉進行安裝。

⯈ 26.2 HPA 設定檔範例

前一章我們對 HPA 概念已經有了不少的著墨，但實際上在網路上關於 HPA 的 API 設定檔，對於其他資源來說可以說相對少很多，所以這裡整理了大致上會使用到的一些實用參數。

```
1.   apiVersion: autoscaling/v2
2.   kind: HorizontalPodAutoscaler
3.   metadata:
4.     name: php-apache
5.   spec:
6.     scaleTargetRef:
7.       apiVersion: apps/v1
8.       kind: Deployment
9.       name: php-apache
10.    minReplicas: 1
11.    maxReplicas: 10
12.    metrics:
13.    - type: Resource
14.      resource:
15.        name: cpu
16.        target:
17.          type: Utilization
18.          averageUtilization: 50
19.    - type: Resource
20.      resource:
```

```
21.          name: memory
22.          target:
23.            type: Utilization
24.            averageUtilization: 80
25.   - type: Pods
26.     pods:
27.       metric:
28.         name: packets-per-second
29.       target:
30.         type: AverageValue
31.         averageValue: 1k
32.   - type: Object
33.     object:
34.       metric:
35.         name: requests-per-second
36.       describedObject:
37.         apiVersion: networking.k8s.io/v1beta1
38.         kind: Ingress
39.         name: main-route
40.       target:
41.         type: Value
42.         value: 10k
43.   - type: External
44.     external:
45.       metric:
46.         name: queue_messages_ready
47.         selector:
48.           matchLabels:
49.             queue: "worker_tasks"
50.       target:
51.         type: AverageValue
52.         averageValue: 30
53.   behavior:
```

```
54.      scaleDown:
55.        stabilizationWindowSeconds: 300
56.        policies:
57.        - type: Percent
58.          value: 100
59.          periodSeconds: 15
60.      scaleUp:
61.        stabilizationWindowSeconds: 0
62.        policies:
63.        - type: Percent
64.          value: 100
65.          periodSeconds: 15
66.        - type: Pods
67.          value: 4
68.          periodSeconds: 15
69.        selectPolicy: Max
```

一些重要的設定參數：

- apiVersion：autoscaling / v2beta2 後開始可以使用 Metrics Server 中的 memory 當作擴縮指標。
- spec.minReplicas / maxReplicas：定義最小或最大的 replica，不能設定 0。
- spec.metrics：定義需要監控的資源，及其搭配的使用量調整。
  - type：
    - resource：resource 是指 Kubernetes 已知的資源指標，此結構描述當前擴縮目標中的每個 Pod 的 CPU 或記憶體。
    - pods：pods 是指當前擴縮目標的每個 Pod 的指標，再與目標值進行比較前，這些指標數值將被平均。
      - metric.selector：selector 可以直接看作是 labelSelector，可以用來指定需要獲取的具體指標範圍。未設置時，僅以 metricName 參數用於收集資源指標。

- ▸ object：object 是指單個 Kubenetes 物件的指標，例如範例中的 Ingress 。
  - metrics.object.describedObject：describeObject 中指定足夠的資訊，辨別所引用的資源。
  - metric.selector：此 selector 跟上述的用法相同，用以約束更明確的資源指標收集範圍。
- ▸ external：external 是指以非 Kubernetes 資源物件當作擴展指標依據，外部指標使我們可以使用來自監控系統的任何指標來自動擴展叢集。你需要在 metric 提供 name 和 selector。

- {resource/pods/object/external}.(resource).target：
  - type：定義特定指標的目標值、平均值或平均使用率，type 表示指標類別是 Utilization、Value 或 AverageValue。
  - averageUtilization：averageUtilization 是跨所有相關 Pod 得出的資源指標均值的目標值，表示為 Pod 資源請求值的百分比。目前僅對「Resource」指標源類別有效。
  - averageValue：averageValue 是跨所有 Pod 得出的指標均值的目標值（以數量形式給出）。
  - value：value 是指標的目標值（以數量形式給出）。

- behavior：官方預設值，依照實際場景調整。
  - stabilizationWindowSeconds：主要目的是為了減少 HPA 的反應速度，使其在特定的時間窗口內不會對瞬時的負載波動做出劇烈的調整。具體來說，stabilizationWindowSeconds 定義了一個時間窗口，在該窗口期間，HPA 會考慮最不利於當前操作的指標值（當需要擴展時，它會選擇最大值，而當需要縮減時，它會選擇最小值）。
  - {scaleUp/scaleDown}.policy：代表每隔一段 periodSeconds 的時間，副本數變化最多不會超過 Percent 或 Pods 定義的數量。

⫸ 26.3 實戰演練

接下來我們就用官方範例來探討一下細節。

範例檔 **deployment.yaml**

```
1.  # deployment.yaml
2.  apiVersion: apps/v1
3.  kind: Deployment
4.  metadata:
5.    name: php-apache
6.  spec:
7.    selector:
8.      matchLabels:
9.        run: php-apache
10.   replicas: 1
11.   template:
12.     metadata:
13.       labels:
14.         run: php-apache
15.     spec:
16.       containers:
17.         - name: php-apache
18.           image: registry.k8s.io/hpa-example
19.           ports:
20.             - containerPort: 80
21.           resources:
22.             limits:
23.               cpu: 500m
24.               memory: 512Mi
25.             requests:
26.               cpu: 500m
```

```
27.              memory: 512Mi
28. ---
29. apiVersion: v1
30. kind: Service
31. metadata:
32.   name: php-apache
33.   labels:
34.     run: php-apache
35. spec:
36.   ports:
37.     - port: 80
38.   selector:
39.     run: php-apache
```

範例檔　**hpa.yaml**

```
1.  # hpa.yaml
2.  apiVersion: autoscaling/v2
3.  kind: HorizontalPodAutoscaler
4.  metadata:
5.    name: php-apache
6.  spec:
7.    scaleTargetRef:
8.      apiVersion: apps/v1
9.      kind: Deployment
10.     name: php-apache
11.   minReplicas: 1
12.   maxReplicas: 10
13.   metrics:
14.     - type: Resource
15.       resource:
16.         name: cpu
17.         target:
```

```
18.          type: Utilization
19.          averageUtilization: 50
20.     - type: Resource
21.       resource:
22.         name: memory
23.         target:
24.           type: Utilization
25.           averageUtilization: 80
```

來建立起資源吧：

```
1.  kubectl apply -f ./deployment.yaml
2.  kubectl apply -f ./hpa.yaml
```

接下來我們將運行個簡單壓力測試指令查看 Pod 的 HPA 狀況：

```
1.  kubectl run -i --tty load-generator --rm --image=busybox: 1.28
    --restart=Never -- /bin/sh -c "while sleep 0.01; do wget -q -O-
    http://php-apache; done"
2.  --------
3.  If you don't see a command prompt, try pressing enter.
4.  OK!OK!OK!OK!OK!OK!OK!OK!OK!OK!OK!OK!OK!OK!OK!OK!OK!OK!
```

執行 HPA 監控：

```
1.  kubectl get hpa --watch
2.  ------
3.  NAME           REFERENCE              TARGETS
    MINPODS    MAXPODS    REPLICAS    AGE
4.  php-apache    Deployment/php-apache    0%/80%, <unknown>/50%
    1          10         1           3s
5.  php-apache    Deployment/php-apache    1%/80%, 0%/50%
    1          10         1           33s
6.  php-apache    Deployment/php-apache    1%/80%, 0%/50%
```

```
     1          10          1          48s
7.   php-apache  Deployment/php-apache   1%/80%, 98%/50%
     1          10          3          63s
8.   php-apache  Deployment/php-apache   2%/80%, 29%/50%
     1          10          3          94s
9.   php-apache  Deployment/php-apache   2%/80%, 33%/50%
     1          10          3          2m4s
```

成功看到系統成功水平擴展並且分散服務負擔！

筆者碎碎念

經過對 HPA 的初步探索與練習後，我們對其運作原理與應用範疇有了更深一層的理解。

Kubernetes 的 HPA 雖然提供了基本的容器擴縮功能，但在面對真正的大流量正式環境，例如某些熱點新聞的高流量情況下，可能在短短幾分鐘內需要將機器數量擴大數倍，此時就顯得有些力不從心。在這種情況下，更細緻的設定以及對更多外部資源指標的依賴是必要的。而這正是像 Grafana 和 Prometheus 這種監控工具的價值所在。這種組合可以提供更精細、更靈活的監控和擴縮策略，能夠更好地應對突然的流量激增。無論如何，我們需要持續學習與嘗試，以便能更好地應對各種可能的情況。

參考資料

- Horizontal Pod Autoscaling
  https://kubernetes.io/docs/tasks/run-application/horizontal-pod-autoscale/

- HorizontalPodAutoscaler Walkthrough
  https://kubernetes.io/docs/tasks/run-application/horizontal-pod-

autoscale-walkthrough/ dashboard k8s 查看 hpa

- 從異世界歸來的第二六天 - Kubernetes AutoScaling(二) - Horizontal Pod Autoscaler
  https://www.notion.so/Day21-Kubernetes-AutoScaling-Horizontal-Pod-Autoscaler-8429203e2f7b4b16a91d6e08585277a9

- HorizontalPodAutoscaler API
  https://kubernetes.io/zh-cn/docs/reference/kubernetes-api/workload-resources/horizontal-pod-autoscaler-v2/

- Kubernetes Horizontal Scaling/Vertical Scaling 概念
  https://sean22492249.medium.com/kubernetes-horizontal-scaling-vertical-scaling-%E6%A6%82%E5%BF%B5-e8e70ce6f034

Kubernetes AutoScaling — Vertical Pod AutoScaler

在先前的章節，我們已深入探討了 HPA 的水平擴展功能。接下來，我們將著重介紹 VPA 的垂直擴展能力。當我們首次部署一項新的服務時，通常難以確定應為該服務設定多少資源。要達到理想的資源設定，我們可能得耗費大量時間進行手動監控和調整。若存在一種機制能根據實際使用情況提供客觀建議，或者自動地調整資源設定，那將極大地助益我們，而這正是 VPA 所能做到的。然而，儘管 VPA 已推出超過兩年，我們在 Kubernetes 的官方文件中仍然找不到完整的介紹或使用範例。因此，在這一章中，我們會融合多種資源，為大家詳盡介紹 VPA 的各項功能和細節。

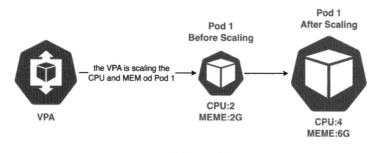

▲ 圖 27-1 VPA

27.1 確認 Metrics Server 是否就緒

在進行之前，我們需要擁有一個已設置 Metrics Server 的 Kubernetes 叢集來收集各種資源指標，當作 autoscaling 的依據。

```
1.  kubectl top node
2.  -------
3.  NAME            CPU(cores)    CPU%    MEMORY(bytes)    MEMORY%
4.  docker-desktop  258m          6%      5717Mi           72%
```

如果還沒有安裝的朋友可以參考 24.3 節〈安裝 Metrics Server〉進行安裝。

⊪ 27.2 VPA 元件以及運作流程

VPA 的運作依賴於三個主要的元件：

- Recommender：該元件負責監控資源使用情況並進行預估計算。它查看歷史指標並根據這些資料調整請求和限制設定。
- Updater：此元件負責驅逐需要更新的 Pod。這是因為對 Request / Limit 的更新需要重啟服務。如果已設定 updateMode: Auto，那麼 Recommender 的任何推薦都將觸發 Updater 更新 Pod。
- Admission Controller：一旦 Updater 驅逐了 Pod，並且在資源控制器重新建立 Pod 之前，將透過 Webhook 觸發 Admission Controller，從而更新 Request / Limit。

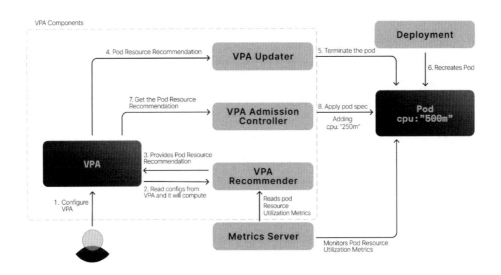

▲ 圖 27-2 VPA Arch

從圖 27-2 我們可以清楚的理解三個元件彼此互動的模式。

27.3 安裝 Custom Resource — VPA

由於官方內建的 API 只有支援 HPA 而已，所以如果我們需要使用到 cluster autoscaler 或者是 vertical autoscaler 等 CRD 時，需要以模組的方式載入。

下載官方 autoscaler repo 並進入 VPA 檔案路徑：

```
1.  git clone git@github.com:kubernetes/autoscaler.git
2.  cd ./autoscaler/vertical-pod-autoscaler
```

執行安裝檔：

```
1.  ./hack/vpa-up.sh
2.  -------
3.  customresourcedefinition.apiextensions.k8s.io/
    verticalpodautoscalercheckpoints.autoscaling.k8s.io created
4.  customresourcedefinition.apiextensions.k8s.io/
    verticalpodautoscalers.autoscaling.k8s.io created
5.  clusterrole.rbac.authorization.k8s.io/system: metrics-reader created
6.  clusterrole.rbac.authorization.k8s.io/system: vpa-actor created
7.  clusterrole.rbac.authorization.k8s.io/system: vpa-checkpoint-actor
    created
8.  clusterrole.rbac.authorization.k8s.io/system: evictioner created
9.  clusterrolebinding.rbac.authorization.k8s.io/system: metrics-reader
    created
10. clusterrolebinding.rbac.authorization.k8s.io/system: vpa-actor created
11. clusterrolebinding.rbac.authorization.k8s.io/system: vpa-
    checkpoint-actor created
12. clusterrole.rbac.authorization.k8s.io/system: vpa-target-reader
    created
13. clusterrolebinding.rbac.authorization.k8s.io/system: vpa-target-
    reader-binding created
```

14. clusterrolebinding.rbac.authorization.k8s.io/system: vpa-evictionter-binding created

15. serviceaccount/vpa-admission-controller created

16. clusterrole.rbac.authorization.k8s.io/system: vpa-admission-controller created

17. clusterrolebinding.rbac.authorization.k8s.io/system: vpa-admission-controller created

18. clusterrole.rbac.authorization.k8s.io/system: vpa-status-reader created

19. clusterrolebinding.rbac.authorization.k8s.io/system: vpa-status-reader-binding created

20. serviceaccount/vpa-updater created

21. deployment.apps/vpa-updater created

22. serviceaccount/vpa-recommender created

23. deployment.apps/vpa-recommender created

24. Generating certs for the VPA Admission Controller in /tmp/vpa-certs.

25. Generating RSA private key, 2048 bit long modulus

26. .+++

27. ....+++

28. ERROR: Failed to create CA certificate for self-signing. If the error is "unknown option -addext", update your openssl version or deploy VPA from the vpa-release-0.8 branch.

29. deployment.apps/vpa-admission-controller created

30. service/vpa-webhook created

這時如果出現了「unknown option -addext」，代表我們需要提升 openssl 版本。

所以我們需要更新 macOS 預設使用的 libressl（openssl 的一個分支）。首先卸載剛剛的安裝：

1. ./hack/vpa-down.sh

使用 brew 更新 libressl：

```
1.  brew install libressl
2.
3.  echo 'export PATH="/opt/homebrew/opt/libressl/bin: $PATH"' >> ~/.zshrc
4.
5.  source ~/.zshrc
```

再次執行安裝檔：

```
1.  ./hack/vpa-up.sh
2.  ------
3.  customresourcedefinition.apiextensions.k8s.io/
    verticalpodautoscalercheckpoints.autoscaling.k8s.io created
4.  customresourcedefinition.apiextensions.k8s.io/
    verticalpodautoscalers.autoscaling.k8s.io created
5.  clusterrole.rbac.authorization.k8s.io/system: metrics-reader created
6.  clusterrole.rbac.authorization.k8s.io/system: vpa-actor created
7.  clusterrole.rbac.authorization.k8s.io/system: vpa-checkpoint-actor
    created
8.  clusterrole.rbac.authorization.k8s.io/system: evictioner created
9.  clusterrolebinding.rbac.authorization.k8s.io/system: metrics-reader
    created
10. clusterrolebinding.rbac.authorization.k8s.io/system: vpa-actor created
11. clusterrolebinding.rbac.authorization.k8s.io/system: vpa-
    checkpoint-actor created
12. clusterrole.rbac.authorization.k8s.io/system: vpa-target-reader
    created
13. clusterrolebinding.rbac.authorization.k8s.io/system: vpa-target-
    reader-binding created
14. clusterrolebinding.rbac.authorization.k8s.io/system: vpa-
    evictionter-binding created
15. serviceaccount/vpa-admission-controller created
```

```
16. clusterrole.rbac.authorization.k8s.io/system: vpa-admission-
    controller created
17. clusterrolebinding.rbac.authorization.k8s.io/system: vpa-admission-
    controller created
18. clusterrole.rbac.authorization.k8s.io/system: vpa-status-reader
    created
19. clusterrolebinding.rbac.authorization.k8s.io/system: vpa-status-
    reader-binding created
20. serviceaccount/vpa-updater created
21. deployment.apps/vpa-updater created
22. serviceaccount/vpa-recommender created
23. deployment.apps/vpa-recommender created
24. Generating certs for the VPA Admission Controller in /tmp/vpa-certs.
25. Generating RSA private key, 2048 bit long modulus
26. ..........+++++
27. ....+++++
28. e is 010001 (0x65537)
29. Generating RSA private key, 2048 bit long modulus
30. .................+++++
31. ...........................................................+++++
32. e is 010001 (0x65537)
33. Signature ok
34. subject=/CN=vpa-webhook.kube-system.svc
35. Getting CA Private Key
36. Uploading certs to the cluster.
37. secret/vpa-tls-certs created
38. Deleting /tmp/vpa-certs.
39. deployment.apps/vpa-admission-controller created
40. service/vpa-webhook created
```

沒有意外的話，剛剛的問題可以順利解決。

查看 VPA 運行狀況：

```
1.  kubectl get pods -n kube-system | grep vpa
2.  -------
3.  vpa-admission-controller-667dd5b58-jsftm    1/1    Running    0    84s
4.  vpa-recommender-5f48d76d7-g7x6m             1/1    Running    0    85s
5.  vpa-updater-6fc5699544-wrvhb                1/1    Running    0    85s
```

查看 VPA 相關資源是否存在：

```
1.  kubectl api-resources | grep vpa
2.  -------
3.  verticalpodautoscalercheckpoints    vpacheckpoint    autoscaling.k8s.
    io/v1                  true         VerticalPodAutoscalerCheckpoint
4.  verticalpodautoscalers              vpa              autoscaling.k8s.
    io/v1                  true         VerticalPodAutoscaler
```

大功告成！

27.4 實戰演練

範例檔 **deployment.yaml**

```
1.  # deployment.yaml
2.  apiVersion: apps/v1
3.  kind: Deployment
4.  metadata:
5.    name: hamster
6.    namespace: default
7.  spec:
8.    selector:
9.      matchLabels:
```

```
10.        app: hamster
11.    replicas: 1
12.    template:
13.      metadata:
14.        labels:
15.          app: hamster
16.      spec:
17.        containers:
18.        - name: hamster
19.          image: k8s.gcr.io/ubuntu-slim:0.1
20.          resources:
21.            requests:
22.              cpu: 100m
23.              memory: 50Mi
24.            limits:
25.              cpu: 2000m
26.              memory: 2Gi
27.          command: ["/bin/sh"]
28.          args:
29.          - "-c"
30.          - "while true; do timeout 0.2s yes >/dev/null; sleep
     0.5s; done"
```

範例檔 **vpa.yaml**

```
1.  # vpa.yaml
2.  apiVersion: autoscaling.k8s.io/v1
3.  kind: VerticalPodAutoscaler
4.  metadata:
5.    name: hamster-vpa
6.    namespace: default
7.  spec:
8.    targetRef:
```

```
9.      apiVersion: apps/v1
10.     kind: Deployment
11.     name: hamster
12.   updatePolicy:
13.     updateMode: "Off"
14.   resourcePolicy:
15.     containerPolicies:
16.       - containerName: '*'
17.         minAllowed:
18.           cpu: 100m
19.           memory: 50Mi
20.         maxAllowed:
21.           cpu: 1
22.           memory: 500Mi
23.         controlledResources: ["cpu", "memory"]
```

來看看相關的參數：

- spec.updatePolic.updateMode：
  - Off：VPA 只會提供推薦資源分配，不會自動的調整任何設定。
  - Initial：VPA 只會在 Pod 被建立時調整資源分配，並且不會再有任何自動調整。
  - Auto：VPA 將會自動設定 Recommender 提供的設定。
  - Recreate：和 Auto 類似差別在於每次重啟 Pod 都會 recreate（很少用到）。
- spec.resourcePolicy.containerPolicies：
  - containerName：指定 VPA 的範圍，「*」代表目標中所有的 Pod。
  - minAllowed：可調整的資源下限。
  - maxAllowed：可調整的資源上限。
  - controlledResources：需要監控的資源指標，有 cpu 和 memory 可以選擇。

執行設定：

```
1.   kubectl apply -f ./deployment.yaml -f ./vpa.yaml
2.   ----------
3.   deployment.apps/hamster created
4.   verticalpodautoscaler.autoscaling.k8s.io/hamster-vpa created
```

查看 VPA 給出的設定建議：

```
1.   kubectl get vpa
2.   ----------
3.   NAME            MODE    CPU     MEM        PROVIDED    AGE
4.   hamster-vpa     Off     379m    262144k    True        2m58s
5.   # 262144K 略等於 255 Mi
```

查看 VPA 的推薦內容：

```
1.   kubectl describe vpa hamster-vpa
2.   ----------
3.   ...
4.   Status:
5.     Conditions:
6.       Last Transition Time:  2022-08-28T10: 03: 36Z
7.       Status:                True
8.       Type:                  RecommendationProvided
9.     Recommendation:
10.     Container Recommendations:
11.        Container Name: hamster
12.        Lower Bound:
13.          Cpu:       204m
14.          Memory:    262144k
15.        Target:
16.          Cpu:       379m
```

```
17.           Memory:   262144k
18.       Uncapped Target:
19.           Cpu:      379m
20.           Memory:   262144k
21.       Upper Bound:
22.           Cpu:      1
23.           Memory:   500Mi
24.  Events:           <none>
```

從 Status.Recommendation 中有幾項重要的數值：

- Lower Bound：如果 Pod 的請求小於下限，則 VPA 會刪除該 Pod 並且將其替換。
- Upper Bound：如果 Pod 的請求大於上限，則 VPA 會刪除該 Pod 並且將其替換。
- Target：該值為在 minAllowed 和 maxAllowed 範圍內的推薦值，旨在使容器以最佳方式運行。
- Uncapped：不受到 minAllowed 和 maxAllowed 範圍限制的推薦值。

∷ 27.5 移除 VPA 模組

```
1.  ./hack/vpa-down.sh
```

再執行一次 vpa-down.sh ，即可將安裝好的客製資源移除。

> **筆者碎碎念**
>
> 我們可以利用各種 autoscaler 節省許多不必要的浪費，更可以更多的結合 HPA 和 VPA，但需要注意的是使用非外部資源指標的 HPA 將會與 VPA 的 Auto 模式互相衝突造成不可預期的問題，所以個人比較偏好使用 HPA 搭配 VPA Off 模式，使用推薦值輔助我們的資源分配。

參考資料

- Kubernetes VPA
  https://www.kubecost.com/kubernetes-autoscaling/kubernetes-vpa/

- Pod 縱向自動擴縮
  https://cloud.google.com/kubernetes-engine/docs/concepts/verticalpodautoscaler

- dashboard k8s 查看 hpa
  https://www.notion.so/Day21-Kubernetes-AutoScaling-Horizontal-Pod-Autoscaler-8429203e2f7b4b16a91d6e08585277a9

- Vertical Pod Autoscaling: Example | Metrics | Limits | Vertical Pod Autoscaler | VPA | Kubernetes
  https://www.youtube.com/watch?v=3h-vDDTZrm8&ab_channel=AntonPutra

- Kubernetes Horizontal Scaling/Vertical Scaling 概念
  https://sean22492249.medium.com/kubernetes-horizontal-scaling-vertical-scaling-%E6%A6%82%E5%BF%B5-e8e70ce6f034

從異世界歸來發現只剩自己不會 Kubernetes
初心者進入雲端世界的實戰攻略！

CHAPTER

28

Kubernetes AutoScaling — Custom Pod AutoScaler

在 Kubernetes 中，原生的 HPA（水平 Pod 自動擴展）功能依據 CPU 和記憶體的平均使用量或使用率作為自動擴展的觸發指標。對於利用 Partition 分配機制的應用程式，例如 Kafka，這種設定可能會引起工作量分布不均，導致資源不能有效伸縮。例如，某些 Pod 的使用率可能接近其負載上限，但由於整體平均使用率未達到閾值，HPA 未能及時進行擴展。在這種情況下，我們需要依靠使用率以外的指標，來判定是否需要進行擴展。

解決這種問題的一個策略是採用 KEDA（Kubernetes Event-Driven Autoscaling）。KEDA 透過監控每個 Pod 的事件消費速度（例如 Kafka 訊息的處理速率）來進行自動擴展，而非僅僅依據全體 Pod 的平均資源使用情況。這種方式對於處理工作量分布不均的問題有很大的幫助，不僅可以更精確地調整 Pod 的數量，也能進一步優化資源使用，提升整體系統效率。

⠿ 28.1 安裝 KEDA

首先需要新增 KEDA 的 HELM Repo：

```
1.  helm repo add kedacore https://kedacore.github.io/charts
```

更新 HELM Repo：

```
1.  helm repo update
```

下載 KEDA：

```
1.  helm install keda kedacore/keda --namespace keda --create-namespace
```

28.2 KEDA（Kubernetes Event-Driven Autoscaling）

KEDA（Kubernetes Event-driven Autoscaling）最初是由 Microsoft 和 Red Hat 共同開發並於 2019 年宣布開源。它的目標是在 Kubernetes 上實現基於事件的自動縮放。

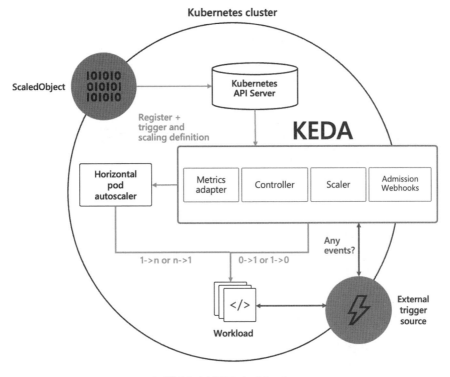

▲ 圖 28-1 KEDA Architecture

Kubernetes 是一個強大的容器管理平台，可以自動調度和管理應用程序。然而，Kubernetes 預設的自動擴展功能僅基於 CPU 使用率等指標，並無法直接根據外部事件來調整應用程序的規模。這對於需要根據事件負載進行彈性擴展的應用程序來說是一個挑戰。

為了解決這個問題，KEDA 在 Kubernetes 中引入了事件驅動的自動縮放能力。它擴展了 Kubernetes 自動擴展機制，使開發人員能夠根據外部事件（如消息佇列、Kafka、RabbitMQ 等）來調整應用程序的副本數量。KEDA 透過在 Kubernetes Pod 上添加自定義指標，根據事件的數量或其他屬性來縮放應用程序，從而實現了基於事件的自動擴展。

KEDA 的目標是提供一個通用、彈性和可擴展的事件驅動自動擴展解決方案，以幫助開發人員更好地在 Kubernetes 環境中管理和優化它們的應用程序。它已經得到了廣泛社群的支持和貢獻，並且在 Kubernetes 社區中越來越受歡迎。KEDA 已經成為 CNCF（Cloud Native Computing Foundation）的畢業項目，這進一步表明了它在容器和雲端原生領域的重要性和影響力。

28.3 KEDA CRD — ScaledObject 和 ScaledJob

在 KEDA（Kubernetes Event-Driven Autoscaling）的框架中，ScaledObject 和 ScaledJob 是兩種定義自動伸縮行為的自定義資源（Custom Resource Definitions，CRDs），這兩種 CRDs 提供了一個彈性且詳細的方式來設定如何擴展和縮小。

1. ScaledObject

ScaledObject 是用於描述如何自動伸縮 Deployment 或 StatefulSet 的資源。它定義了要監控的觸發器類型（例如 Kafka、RabbitMQ、Prometheus metrics 等）、觸發器的詳細資訊以及如何根據這些觸發器自動伸縮的規則。

```
1.  apiVersion: keda.sh/v1alpha1
2.  kind: ScaledObject
3.  metadata:
4.    name: my-scaled-object
```

```
5.      namespace: default
6.   spec:
7.      scaleTargetRef:
8.        name: my-deployment
9.      minReplicaCount: 0
10.     maxReplicaCount: 100
11.     triggers:
12.     - type: kafka
13.       metadata:
14.         # Required
15.         bootstrapServers: my-cluster-kafka-bootstrap.kafka: 9092
16.         consumerGroup: my-group
17.         topic: my-topic
```

在此範例中，一個 Kafka 事件將觸發 KEDA 來自動伸縮名為 my-deployment 的
Deployment。

通用參數：

- scaleTargetRef：這部分用來參照 ScaledObject 所應用的資源。這是 KEDA
  會依據 triggers 來進行擴縮並設定 HPA 的資源。

- pollingInterval：這是 KEDA 檢查每個觸發器的間隔。預設情況下，KEDA 會
  每隔 30 秒檢查一次每個 ScaledObject 的觸發源。

- cooldownPeriod：觸發發生後，KEDA 在將資源縮放回 0 前等待的時間。預
  設情況下是 5 分鐘（300 秒）。

- idleReplicaCount：如果設定了這個屬性，KEDA 會將資源縮放到這個副本數
  量。這個設定必須少於 minReplicaCount。

- minReplicaCount：KEDA 將資源縮放到的最小副本數。預設情況下為 0。

- maxReplicaCount：此設定被傳遞給 KEDA 為給定資源建立的 HPA 定義，並
  保存目標資源的最大副本數。

- fallback：可選的。定義了當伸縮器處於錯誤狀態時，需要回溯的副本數量。

- advanced：可選的。在這裡可以設定一些進階選項，例如 restoreToOriginal

ReplicaCount 和 horizontalPodAutoscalerConfig。

- triggers：觸發器列表，用來啟動對目標資源的縮放。

2. ScaledJob

ScaledJob 類似於 ScaledObject，但它是用來自動伸縮 Kubernetes Jobs 的。
一旦觸發器被觸發，KEDA 將建立一個或多個 Job 實例來處理該事件。

```
1.  apiVersion: keda.sh/v1alpha1
2.  kind: ScaledJob
3.  metadata:
4.    name: rabbitmq-consumer
5.    namespace: default
6.  spec:
7.    jobTargetRef:
8.      template:
9.        spec:
10.         containers:
11.         - name: rabbitmq-client
12.           image: rabbitmq-client: latest
13.           command:
14.           - receive
15.           args:
16.           - "$(RABBITMQ_QUEUE)"
17.   pollingInterval: 30
18.   successfulJobsHistoryLimit: 5
19.   failedJobsHistoryLimit: 5
20.   maxReplicaCount: 30
21.   triggers:
22.   - type: rabbitmq
23.     metadata:
24.       queueName: hello
25.       host: RabbitMqHost
26.       queueLength: '5'
```

在這個例子中，一旦 RabbitMQ 佇列中有新的消息，KEDA 將自動建立新的 Kubernetes Job 實例來處理這些消息。

通用參數：

- jobTargetRef：描述作業模板。包括如下參數：
  - parallelism：最大期望實例數。
  - completions：完成的期望實例數。
  - activeDeadlineSeconds：從作業開始時間起，作業可以在系統試圖終止它之前保持活動的秒數。
  - backoffLimit：標記此作業失敗前的重試次數。
- pollingInterval：檢查每個觸發器的間隔時間。
- successfulJobsHistoryLimit 和 failedJobsHistoryLimit：保存的已完成作業和失敗作業的數量。
- envSourceContainerName：KEDA 應試圖獲取其中包含秘密等的環境屬性的作業中的容器名稱。
- minReplicaCount/maxReplicaCount：預設建立的作業數量的最小值和最大值。
- rollout：描述了 KEDA 更新現有 ScaledJob 時將使用的滾動策略。
- scalingStrategy：選擇使用的縮放策略，可能的值為 default、custom 或 accurate。

⫸ 28.4 KEDA 觸發器

KEDA 目前支援多達六十多種的 Event Source 作為觸發擴展收縮的資料來源，這讓我們有極高的彈性去選擇最適合你應用程式的觸發條件。更驚人的是，即使在這六十多種 Event Source 中仍找不到適合的，KEDA 也提供了簡單易用的擴展機制，可以快速將自定義的指標整合入 KEDA 的觸發機制中。接下來我們將介紹一些常見的 Scaler 種類，並透過它們理解 KEDA 的強大之處：

1. Prometheus：這是一個開源的監控與警報工具，專為處理指標資料而設計。在 KEDA 中，我們可以使用 Prometheus Scaler 來偵測單一 Pod 的使用率，並根據此資料做出擴展的動作。

```
1.    apiVersion: keda.sh/v1alpha1
2.    kind: ScaledObject
3.    metadata:
4.      name: prometheus-scaledobject
5.      namespace: default
6.    spec:
7.      scaleTargetRef:
8.        name: my-deployment
9.      triggers:
10.     - type: prometheus
11.       metadata:
12.         serverAddress: http://<prometheus-host>: 9090
13.         metricName: http_requests_total
14.         threshold: '100'
15.         query: sum(rate(http_requests_total{deployment="my-
   deployment"}[2m]))
16.
```

2. Kafka：Kafka 是一個分散式的資料串流平台，可以處理實時資料串流。KEDA 的 Kafka Scaler 可以根據 Kafka 的訊息消耗延遲數量來作為是否需要擴展資源的依據。

```
1.    apiVersion: keda.sh/v1alpha1
2.    kind: ScaledObject
3.    metadata:
4.      name: kafka-scaledobject
5.      namespace: default
6.    spec:
7.      scaleTargetRef:
```

```
8.          name: my-deployment
9.        pollingInterval: 30
10.       triggers:
11.       - type: kafka
12.         metadata:
13.           bootstrapServers: localhost: 9092
14.           consumerGroup: my-group
15.           topic: test-topic
16.           # Optional
17.           lagThreshold: "50"
18.           offsetResetPolicy: latest
```

3. PostgreSQL：在 KEDA 中，可以利用 PostgreSQL Scaler 根據 PostgreSQL 靈活地查詢到目前任務數量，以決定是否需要調節資源。

```
1.        apiVersion: keda.sh/v1alpha1
2.        kind: ScaledObject
3.        metadata:
4.          name: airflow-worker
5.        spec:
6.          scaleTargetRef:
7.            name: airflow-worker
8.          pollingInterval: 10    # Optional. Default: 30 seconds
9.          cooldownPeriod: 30     # Optional. Default: 300 seconds
10.         maxReplicaCount: 10    # Optional. Default: 100
11.         triggers:
12.           - type: postgresql
13.             metadata:
14.               connectionFromEnv: AIRFLOW_CONN_AIRFLOW_DB
15.               query: "SELECT ceil(COUNT(*): : decimal / 16) FROM
      task_instance WHERE state='running' OR state='queued'"
16.               targetQueryValue: 1
```

4. Google Cloud Platform Storage：不僅本地資源，KEDA 甚至可以利用雲端
   平台如 Google Cloud Platform 的儲存服務作為觸發指標。這意味著我們的
   應用程式可以根據雲端儲存的狀態進行自動擴展。

```
1.    apiVersion: keda.sh/v1alpha1
2.    kind: TriggerAuthentication
3.    metadata:
4.      name: keda-trigger-auth-gcp-credentials
5.    spec:
6.      secretTargetRef:
7.      - parameter: GoogleApplicationCredentials
8.        name: gcp-storage-secret        # Required. Refers to the
   name of the secret
9.        key: GOOGLE_APPLICATION_CREDENTIALS_JSON        # Required.
10.   ---
11.   apiVersion: keda.sh/v1alpha1
12.   kind: ScaledObject
13.   metadata:
14.     name: gcp-storage-scaledobject
15.   spec:
16.     scaleTargetRef:
17.       name: keda-gcp-storage-go
18.     triggers:
19.     - type: gcp-storage
20.       authenticationRef:
21.         name: keda-trigger-auth-gcp-credentials
22.       metadata:
23.         bucketName: "Transactions"
24.         targetObjectCount: "5"
25.         blobPrefix: blobsubpath # Default : ""
26.         blobDelimiter: "/"
```

5. Elasticsearch：相信不少公司是以 Elasticsearch 作為 Log、Metrics 等資料的
   解決方案，藉著優異的查詢效能以及 Kibana 豐富的視覺化介面，現在我們
   也能直接以 Elasticsearch 作為我們跨維度的擴展指標了。

```
1.  apiVersion: v1
2.  kind: Secret
3.  metadata:
4.    name: elasticsearch-secrets
5.  type: Opaque
6.  data:
7.    password: cGFzc3cwcmQh
8.  ---
9.  apiVersion: keda.sh/v1alpha1
10. kind: TriggerAuthentication
11. metadata:
12.   name: keda-trigger-auth-elasticsearch-secret
13. spec:
14.   secretTargetRef:
15.   - parameter: password
16.     name: elasticsearch-secrets
17.     key: password
18. ---
19. apiVersion: keda.sh/v1alpha1
20. kind: ScaledObject
21. metadata:
22.   name: elasticsearch-scaledobject
23. spec:
24.   scaleTargetRef:
25.     name: "deployment-name"
26.   triggers:
27.   - type: elasticsearch
28.     metadata:
29.       addresses: "http://localhost: 9200"
30.       username: "elastic"
```

```
31.          index: "my-index"
32.          searchTemplateName: "my-search-template"
33.          valueLocation: "hits.total.value"
34.          targetValue: "10"
35.          params: "dummy_value: 1"
36.     authenticationRef:
37.          name: keda-trigger-auth-elasticsearch-secret
```

28.5 KEDA 中的防抖動機制 Debouncing

在 KEDA（Kubernetes Event-Driven Autoscaling）系統中，使用了一套特定
的參數設定來實現防抖動的機制，以確保系統不會因為瞬間的負載變化而過
度擴展或收縮。這套防抖動機制的實現，是利用了 KEDA 的兩個重要參數：
cooldownPeriod 和 pollingInterval，並且結合了 Kubernetes 原生的 Horizontal
Pod Autoscaler（HPA）中的 behavior 參數設定來達成。讓我們一起看看這些參
數是如何一起工作，建立起一個穩定的自動縮放系統：

1. cooldownPeriod：這是在 KEDA 觸發一次擴展或縮放事件後，將系統資源
   縮放回原始狀態之前需要等待的時間。預設情況下這個時間是 5 分鐘（300
   秒）。請注意，在 KEDA 將系統從多副本縮放回到 0 副本的情況下，這個時
   間間隔才會生效。然而，如果系統從 1 副本縮放到更多副本時，這個行為則
   是由 Kubernetes 的 HPA 控制的，而不是由 KEDA 的 cooldownPeriod 控制。

2. pollingInterval：這是 KEDA 決定多久檢查一次新的指標資料，以確定是否
   需要進行縮放操作的時間間隔。這個參數的設定對於 KEDA 反應指標變化的
   速度有直接的影響。

3. behavior：這是原生 HPA 中用來控制系統擴展和收縮行為的參數設定。你可
   以透過調整這個設定來更細緻地控制你的系統縮放行為，相關詳細設定可以
   參考之前的 26.2 節。

接下來,讓我們看一下在 KEDA 的 ScaledObject 設定中,這些參數是如何被使用的:

```
1.   apiVersion: keda.sh/v1alpha1
2.   kind: ScaledObject
3.   metadata:
4.     name: prometheus-scaledobject
5.     namespace: default
6.   spec:
7.     scaleTargetRef:
8.       name: prometheus-example-usage
9.     pollingInterval: 15  # Optional. Default: 30 seconds
10.    cooldownPeriod:  300 # Optional. Default: 300 seconds
11.    minReplicaCount: 0   # Optional. Default: 0
12.    maxReplicaCount: 100 # Optional. Default: 100
13.    advanced:
14.      horizontalPodAutoscalerConfig:
15.        behavior:
16.          scaleDown:
17.            stabilizationWindowSeconds: 300
18.            policies:
19.            - type: Percent
20.              value: 100
21.              periodSeconds: 15
22.    triggers:
23.    - type: prometheus
24.      metadata:
25.        serverAddress: http://prometheus.prom.svc.cluster.local: 9090
26.        metricName: http_requests_total
27.        threshold: '1'
28.        query: sum(rate(http_requests_total{app='my-app'}[2m]))
```

在這個設定中,我們可以看到 pollingInterval 已經被設定為每 15 秒檢查一次是否需要進行縮放操作,而 cooldownPeriod 則被設定為 300 秒。這表示一旦觸發了一次擴展或收縮事件，系統將會等待 5 分鐘才將資源縮放回原狀。
而在 horizontalPodAutoscalerConfig 的設定中，我們可以看到 scaleDown 的
stabilizationWindowSeconds 也被設定為 300 秒，這表示系統在進行縮放操
作時，會有一個 5 分鐘的穩定窗口。而 policies 的 periodSeconds 設定為 15
秒，這表示每 15 秒允許一次的縮放操作。

28.6 超越 Kubernetes HPA 的彈性伸縮

Kubernetes Event-Driven Autoscaling（KEDA）擁有獨特的能力，能在
Kubernetes 中實現事件驅動的自動縮放，特別是從零到一以及從一到零的縮
放，這是原生的 Kubernetes Horizontal Pod Autoscaler（HPA）無法實現的。

- 0 到 1 的縮放：KEDA 在 Kubernetes 環境中，扮演一個客製化的指標伺服器
  角色，為 Kubernetes HPA 提供了一個抽象層，讓其能夠明瞭目前的事件來
  源負載。KEDA 將會不斷地檢查這些指標，以決定是否需要對服務進行擴展
  或縮小。
  當某個事件來源（例如 Kafka 主題或 RabbitMQ 佇列）中的訊息數量超過預
  設門檻時，KEDA 會感知到這個變化，並透過 Kubernetes HPA 增加應用程
  序的實例數量。這就是我們所稱的「從零縮放到一」，換句話説，一旦負載
  出現，KEDA 可以將服務的副本數從 0 提升到 1 或更多，以適應需求變化。

- 1 到 0 的縮放：在閒置時間或低負載的情況下，KEDA 也能發揮作用，將
  服務副本數從 1 降低到 0。這是因為，當 KEDA 感知到事件來源（例如，
  Kafka 主題或 RabbitMQ 佇列）的訊息數量降低到門檻以下時，它會透過
  Kubernetes HPA 減少應用程序的實例數量，甚至將其減少到 0，達到「從一
  縮放到零」。這種動態調整的能力使 KEDA 特別適合處理突然增加或不規則
  的工作負載，而不需要手動調整應用程序的副本數量，進而實現了資源的最
  佳化使用，並對系統的性能和成本效益產生積極的影響。

- minReplicaCount：在原生的 Kubernetes HPA 中，minReplicaCount 的最小值為 1，也就是說，當工作負載很低或沒有負載時，應用程式至少需要保持一個副本運行。然而，KEDA 的設計讓 minReplicaCount 的最小值可以設定為 0，這就意味著在無負載的情況下，KEDA 可以將應用程式的副本數量完全縮放至零。這對於需要大量節省資源或只在特定時間有負載的應用程式來說，是一個極大的優勢。

▶ 28.7 激活階段與縮放階段

KEDA 的自動縮放過程分為兩個階段：激活階段[1] 和縮放階段。

1. 激活階段（Activation Phase）：在這個階段，KEDA 需要根據 scaler 的 IsActive 函數的回傳值來決定是否需要將工作負載從零縮放到一，或從一縮放到零。這個階段僅適用於從零縮放到一或從一縮放到零的情況。
2. 縮放階段（Scaling Phase）：在這個階段，KEDA 已經決定將工作負載從零縮放到一，接下來，Kubernetes 的 HPA 控制器將根據從 ScaledObject 中獲得的設定以及 KEDA 提供的指標來決定如何進行縮放。這個階段適用於從一縮放到 N 或從 N 縮放到一的情況。

在 KEDA 中，你可以分別為激活和縮放設定不同的閾值：

- 激活閾值（Activation Thresholds）：這個閾值決定了 scaler 是否應該被激活，並根據此決定判斷是否將工作負載從零縮放至一或從一縮放至零。
- 縮放閾值（Scaling Thresholds）：這個閾值定義了工作負載從一縮放到 N 或從 N 縮放到一的目標值。為了實現這個功能，KEDA 將這個目標值傳給 HPA，然後由 HPA 控制器來處理實際的自動縮放操作。

1　在此文中，筆者選擇使用「激活」來表示，而非「啟動」，這是出於 KEDA 有一個「Activate」階段的考量。在語意上，激活相較於啟動更能表示此行為細節。

如果 minReplicaCount 參數設定為大於或等於 1，那麼這個 scaler 將永遠處於激活狀態，激活閾值將被忽略。

NOTE

請注意，如果 scaler 不支持設定激活閾值（即沒有以「activation」為前綴的屬性），那麼它的激活閾值將始終為 0。如果激活閾值和縮放閾值給出了不同的縮放決策，則激活閾值將有更高的優先級。例如，如果設定為「threshold: 10」和「activationThreshold: 50」，那麼即使 HPA 需要 4 個實例，但只有 40 個訊息，scaler 也不會被激活，並且工作負載會被縮放至零。

筆者碎碎念

隨著雲端計算和微服務架構的發展，事件驅動的自動擴展技術成為一項關鍵技術，其中，Kubernetes Event-Driven Autoscaling（KEDA）表現出了其獨特的優勢。KEDA 不僅能夠實現一到零以及零到一的自動縮放，而且其應用場景遠超過另一套以 Prometheus 為擴縮依據的 Prometheus adaptor。

此外，KEDA 支援的事件源多元且豐富，包括但不限於 Kafka、RabbitMQ、Azure Service Bus 等，對於各種應用都有很好的支援。比起 Prometheus Adaptor，KEDA 提供的更為強大且靈活的自動縮放解決方案，使其有潛力在未來成為主流的 Kubernetes HPA 解決方案。

參考資料

- keda.sh
  https://keda.sh/

Part 9
主題篇—Security
朕不給的，你不能搶

隨著團隊以及產品複雜度的成長，資安以及權限控管的議題出現的次數將會越來越頻繁。

Kubernetes Security — 使用 Context 進行用戶管理

AutoScaling 主題結束後，隨之而來的主題是 Security ，這個議題可以説在
DevOps 中是特別耐人尋味，有趣的點是它不一定是工作中的必備技能，但其
重要性不容忽視，畢竟不是每家公司都有數十人等級以上的團隊，或同時管理
維護大大小小不同產品的工作環境，更多的是三到五人的小開發團隊以及一人
維護的專案，這時候權限管理相關的優先級自然被排到比較後面。

這時我們從另一個角度來看，人人想進的理想好公司通常除了好的文化跟環境
之外，也有很大的概率是一間人數不小規模的企業甚至是跨國集團，如果我們
在職業生涯追求的是這類型的公司，那又有什麼理由不去理解為自己加分的技
能呢？所以我認為不論是否為維運人員，都應該熟悉資安。

29.1 Kubernetes 的認證與授權

不論我們藉由指令或是其他方式與我們的 Kubernetes 叢集互動時，一定都
會經過該叢集管理所有物件資源的入口 kube-apiserver ，這時就可以明白
kube-apiserver 身份驗證的重要性，它能替我們檢查所有進入的請求，包括
Authentication（身份驗證）、Authorization（授權）和 Admission Control（許
可控制）。

Kubernetes API 請求從發起到持久化到 ETCD 資料庫中的過程如圖 29-1 所示：

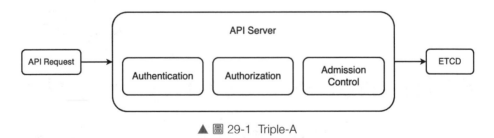

▲ 圖 29-1 Triple-A

接下來我們使用 Kubernetes Context 實戰我們在工作中會面對到的用戶管理認證（Authentication）問題。

▶ 29.2 Kubernetes Context 是什麼？

在 Kubernetes 中，一個 Context 可被視為在客戶端記錄的 Alias，方便且好讀，用於指明如何與叢集溝通。當我們切換到某個叢集後，kubectl 送出的每個指令都會指向該 Context 所設定的 cluster、namespace、user 執行操作。一個關鍵的重點是，kube-apiserver 並不理解所謂的「Context」。而在請求被送出前，客戶端會先將相關的設定轉化為參數。這也與我們先前提到的「Alias」概念相呼應。

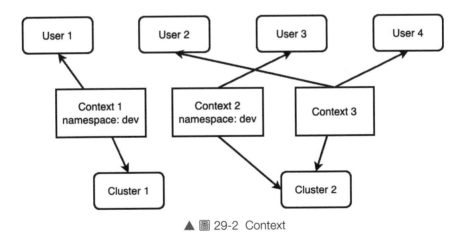

▲ 圖 29-2 Context

先來看看 kubeconfig 中記錄的 Context 到底長什麼樣：

```
1.   kubectl config view
2.   ---------
3.
4.   apiVersion: v1ig view
```

```
5.  clusters:
6.  - cluster:
7.      certificate-authority-data: DATA+OMITTED
8.      server: https://kubernetes.docker.internal: 6443
9.    name: docker-desktop
10. contexts:
11. - context:
12.     cluster: docker-desktop
13.     user: docker-desktop
14.   name: docker-desktop
15. current-context: docker-desktop
16. kind: Config
17. preferences: {}
18. users:
19. - name: docker-desktop
20.   user:
21.     client-certificate-data: REDACTED
22.     client-key-data: REDACTED
```

只查看當前的 context config 資訊：

```
1.  kubectl config view --minify
```

基本上可以歸納出三個重點：

- Clusters：此列表中記錄著每個叢集如何與該叢集的 kube-apiserver 溝通的 URL 以及認證權限。
- Contexts：此列表中記錄著在特定 Context 下，對 kube-apiserver 溝通時將會以內容中的 cluster、user、namespace 來執行（例如 namespcae 為空則預設為 default）。
- Users：此選項定義了每個使用者的唯一名稱和相關的認證授權資訊，像是 client certificates、bearer tokens、authenticating proxy 等等。

⫸ 29.3 用戶管理情境

假設我們的工作環境中有正式環境（Production）和開發環境（Develop）
兩個叢集，而兩個叢集中我們又將前端（Frontend）和後端（Backend）用
namespace 區隔開來，接著我們將創造出前端開發者以及後端開發者兩種角
色，並使用 Context 確保他們只能在相對應的 namespace 控制叢集，不論是在
正式或開發環境。

此時粗略的 kubeconfig 大概如下：

```
1.  apiVersion: v1
2.  kind: Config
3.
4.  clusters:
5.  - cluster:
6.    name: development
7.  - cluster:
8.    name: production
9.
10. users:
11. - name: backend-developer
12. - name: front-developer
13.
14. contexts:
15. - context:
16.   name: dev-backend
17.       namespace: backend
18.       cluster: development
19.       user: backend-developer
20. - context:
21.   name: prod-backend
22.       namespace: backend
```

```
23.        cluster: production
24.        user: backend-developer
25. - context:
26.   name: dev-frontend
27.        namespace: frontend
28.        cluster: development
29.        user: frontend-developer
30. - context:
31.   name: prod-frontend
32.        namespace: frontend
33.        cluster: production
34.        user: frontend-developer
```

當我們切換到某個 Context 時，會以該 user 的認證身份在指定的 cluster 中，向 kube-apiserver 發出對該 namespace 資源的請求，即代表著該 user 在此 Context 下所擁有的操作權限。

▶ 29.4 實戰演練

首先建立出所有的 Context：

1. 建立 / 修改 Context 指令

```
1.  kubectl config set-context <CONTEXT_NAME> --namespace=<NAMESPACE_
    NAME>--cluster=<CLUSTER_NAME> --user=<USER_NAME>
```

2. 陸續將所有 Context 建立出來

```
1.  kubectl config set-context prod-frontend --cluster=production
    --namespace=frontend --user=prod-frontend
2.  --------
```

```
3.  Context "prod-frontend" created.
4.
5.  kubectl config set-context prod-backend --cluster=production
    --namespace=backend --user=prod-backend
6.  --------
7.  Context "prod-backend" created.
8.
9.  kubectl config set-context dev-frontend --cluster=development
    --namespace=frontend --user=dev-frontend
10. --------
11. Context "dev-frontend" created.
12.
13. kubectl config set-context dev-backend --cluster=development
    --namespace=backend --user=dev-backend
14. --------
15. Context "dev-backend" created.
```

3. 再次查看 kubeconfig 可以發現上列 Context 以出現在列表之中

```
1.  kubectl config view
2.  -------
3.  contexts:
4.  - context:
5.      cluster: development
6.      namespace: backend
7.      user: dev-backend
8.    name: dev-backend
9.  - context:
10.     cluster: development
11.     namespace: frontend
12.     user: dev-frontend
13.   name: dev-frontend
14. - context:
```

```
15.     cluster: docker-desktop
16.      user: docker-desktop
17.    name: docker-desktop
18. - context:
19.      cluster: production
20.      namespace: backend
21.      user: prod-backend
22.    name: prod-backend
23. - context:
24.      cluster: production
25.      namespace: frontend
26.      user: prod-frontend
27.    name: prod-frontend
```

4. 切換到所需的 Context

```
1.   # 顯示當前使用的 context
2.   kubectl config current-context
3.   # 切換到指定的 context
4.   kubectl config use-context <CONTEXT_NAME>
```

此時我們已經可以隨心所欲的設定並切換 Context ，在團隊中只要讓對應的角色設定到正確的 Context ，即可實現讓該 user 在指定 cluster 中執行被賦予的權限。

5. 刪除 config 資源

```
1.   # 刪除用戶
2.   kubectl config unset users.<name>
3.   # 刪除叢集
4.   kubectl config unset clusters.<name>
5.   # 刪除 context
6.   kubectl config unset contexts.<name>
```

⏵ 29.5 所以說那個 Context 中的 Cluster 跟 User 呢？

看到這裡可能有些敏銳的同學注意到了，既然 Context 如前面所說的是一種 Alias 的概念，那實際執行的 cluster 授權以及 user 認證又是如何確定的呢？

由於我們在本地使用的 docker-desktop 只能提供我們一個 cluster，加上在實際工作中使用的大多是雲端平台整合好的 Kubernetes 服務（例如 Google GKE，AWS EKS……等），各家平台對於 kubeconfig 的 cluster 管理多半有與自家用戶權限整合，並以自家 cli 或 sdk 的方式進行操作，各家操作方式不一，所以本章重點將放在 Context 管理上。

簡單使用 Google 的 gcloud auth 相關指令新增了對 Google GKE 叢集的 kubeconfig，大概會如下所示：

```
1.   kubectl config view
2.   --------
3.   apiVersion: v1
4.   clusters:
5.   - cluster:
6.       certificate-authority-data: DATA+OMITTED
7.       server: https://35.xxx.xxx.xxx
8.     name: gke_xxx-xxxx_asia-east1-c_xxx-dev
9.   contexts:
10.  - context:
11.      cluster: gke_xxx-xxx_asia-east1-c_xxx-dev
12.      namespace: rtmp-relay
13.      user: gke_xxx-xxx_asia-east1-c_xxx-dev
14.    name: gke_xxx-xxx_asia-east1-c_xxx-dev
15.  current-context: docker-desktop
16.  kind: Config
```

```
17. preferences: {}
18. users:
19. - name: gke_xxx-xxx_asia-east1-c_xxx-dev
20.   user:
21.     auth-provider:
22.       config:
23.         access-token:
24.         cmd-args: config config-helper --format=json
25.         cmd-path: /google-cloud-sdk/bin/gcloud
26.         expiry: "2022-09-23T17: 43: 59Z"
27.         expiry-key: '{.credential.token_expiry}'
28.         token-key: '{.credential.access_token}'
29.       name: gcp
```

關於使用 kubectl 獨自建立 / 修改 cluster 的指令：

```
1.  kubectl config set-cluster \
2.  <CLUSTER_NAME> \
3.  --server=<SERVER_ADDRESS> \
4.  --certificate-authority=<CLUSTER_CERTIFICATE>
```

而建立 / 修改 user 的指令也是非常相似的：

```
1.  kubectl config set-credentials \
2.  <USER_NAME> \
3.  --client-certificate=<USER_CERTIFICATE> \
4.  --client-key=<USER_KEY>
```

以上指令使用憑證的方式驗證身份，而 --client-certificate 被用來認證 user，--certificate-authority 則用來認證該 cluster。官方提供更多的驗證方式可以參考相關文件 Kubernetes authentication overview。

> **筆者碎碎念**
>
> 在以上的介紹中，我們大概可以了解 Context 這個在很多語言或者工具中，有著不同用途的抽象概念。有了 Context 我們在身為服務守護者的團隊協作中，可以確保每個人該有的最小權限，在個人開發中，也可以防止自己的人為疏忽而造成不可逆轉的損失，接下來我們將更進一步的深入介紹，基於 RBAC 下 Kubernetes 如何對一個角色進行授權。

參考資料

- 配置對多叢集的訪問
  https://kubernetes.io/zh-cn/docs/tasks/access-application-cluster/configure-access-multiple-clusters/

- 基於角色的訪問控制（RBAC）
  https://jimmysong.io/kubernetes-handbook/concepts/rbac.html

- 了解 Kubernetes 中的認證機制
  https://godleon.github.io/blog/Kubernetes/k8s-API-Authentication/

- 管理服務帳號
  https://kubernetes.io/zh-cn/docs/reference/access-authn-authz/service-accounts-admin/

- Kubectl Config Set-Context
  https://www.containiq.com/post/kubectl-config-set-context-tutorial-and-best-practices

- k8s 基於 RBAC 的認證、授權介紹和實踐
  https://iter01.com/657295.html

- Day 19 - 老闆！我可以做什麼：RBAC
  https://ithelp.ithome.com.tw/articles/10195944

- [Day17] k8s 管理篇（三）：User Management、RBAC、Node Maintenance
  https://ithelp.ithome.com.tw/articles/10223717

- 用戶認證
  https://kubernetes.io/zh-cn/docs/reference/access-authn-authz/
  authentication/

Kubernetes Security — RBAC Authorization 授權管理

在前面的 Context 介紹中，我們了解了 Kubernetes 如何管理用戶的身份和權限，但我們並未深入探討其認證機制。這並非因為這個主題不重要，相反的，Kubernetes 中的認證和授權是一門相當深奧且重要的學問，其實現方式和概念都十分複雜。在這篇文章中，我們將深入探討這個主題，尤其是與應用層密切相關的 RBAC（Role-Based Access Control）授權管理機制。

30.1 深入了解 Kubernetes API Server

任何想要取得 Kubernetes 資源的請求，都需要經過 Authentication（身份認證）、Authorization（授權）及 Admission Control（許可控制）的驗證過程，才能依照被授權的權限進行操作。

30.1.1 Authentication 身份認證

Authentication 是為了確認該使用者是否能夠進入 Kubernetes 叢集中，一般可以分為兩種使用者：普通使用者（User）以及服務帳號（Service Account）。

普通使用者（User）：可以理解成，透過 kubectl 指令或 RESTful 請求的身份認證，就可以視為一個普通使用者，而這也是 Context 中列出的 user。

服務帳號（Service Account）：Service Account 本身在 Kubernetes 是屬於 resource 的一種，與普通使用者的全域性不同的是，Service account 是以 namespace 為作用域單位。其針對執行中的 Pod 而言，每個 namespace 被建立時，Kubernetes 都會隨之建立一個名稱為 default 的 Service account，並帶有 token 供未來在此 namespace 中產生的 Pod 使用，所以 Pod 將會依照該 Service account 的 token 與 API Server 進行認證。

而在 Kubernetes 中有幾種驗證方式：

- Certificate
- Token
- OpenID
- Webhook

其中 Certificate 是在普通使用者中被廣泛使用的驗證方式。透過客戶端憑證
進行身份驗證時，客戶端必須先取得一個有效的 X.509 客戶端憑證，然後由
Kubernetes API Server 通過驗證這個憑證來驗證你的身份。當然這個 X.509 憑
證必須由叢集 CA 憑證簽名，看起來有沒有跟 HTTPS 憑證很相似，主要差別在
於 CA 供應方變成 Kubernetes 而已。

NOTE

最後我們將會實作 X.509 憑證完成身份認證，如果對 Certificate 還不夠熟悉的
話，可以參考這篇文章：https://hackmd.io/@yzai/rJXYxFpmq。

30.1.2 Authorization 授權

當我們通過了 Authentication（身份認證）後，僅能代表當前的使用者允許與
Kubernetes API Server 溝通，至於該使用者是否有權限（Permission）請求什
麼資源，就是 Authorization 該登場的時候了，Kubernetes API Server 將會在
這時審查請求的 API 屬性，像是 user、Api request verb、Http request verb、
Resource、Namespace、Api Group……等。

關於 Authorization Mode 有以下幾種模式：

- Node
- ABAC

- RBAC
- Webhook

以下我們會以 RBAC 做個介紹並進行實戰演練。

此外 kubectl 提供 auth can-i 子指令，用於快速查詢 API 審查：

```
1.  # 檢查是否可以對 deployments 執行 create
2.  kubectl auth can-i create deployments --namespace default
3.  -----
4.  Yes
```

30.2 實戰使用 RBAC（Role-Based Access Control）

Role-Based Access Control 顧名思義是指基於 Role 的概念建立的訪問控制，用來調節使用者對 Kubernetes API Server 訪問的方法，在各類大型系統及雲端平台中被廣泛使用。

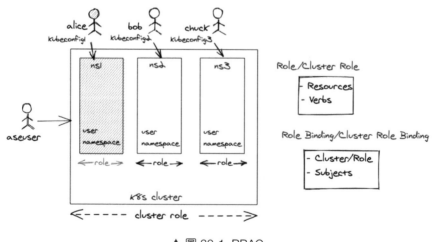

▲ 圖 30-1 RBAC

RBAC 在我們實際工作環境中，以超級管理者為例，這個角色與瀏覽者擁有巨大的權限差距。為了保護內部重要的基礎建設，並不希望每個使用者都可以不受限制的建立、刪除資源，為此我們實現了將權限（Permission）綁定普通使用者以及服務帳號的概念，將複雜的業務權限做的更輕量化，並遵從了權限最小化原則 。

接下來，我們將進行使用者認證，並利用 RBAC 進行角色權限的綁定：

1. 建立 Context 並以 X.509 憑證驗證普通使用者（ user ）

首先我們要以自身 Kubernetes 做為 CA 發送方來產生 CA 憑證。產生一個使用者私鑰：

```
1.  openssl genrsa -out pod-viewer.key 2048
2.  ----
3.  -----BEGIN RSA PRIVATE KEY-----
4.
5.  // ...
6.
7.  -----END RSA PRIVATE KEY-----
```

透過 pod-viewer.key 去產生 CSR（憑證簽名請求），Kubernetes 將會使用憑證中的「subject」的通用名稱（Common Name）欄位來確定使用者名稱：

```
1.  openssl req -new -key pod-viewer.key -out pod-viewer.csr -subj "/
    CN=pod-viewer/O=app"
2.  ------
3.  -----BEGIN CERTIFICATE REQUEST-----
4.
5.  // ...
6.
7.  -----END CERTIFICATE REQUEST-----
```

有了 CSR，我們就可以把它交給叢集管理者（在這裡是指我們）透過叢集 CA 簽署客戶端憑證。

所以我們需要去 docker-desktop 的節點中取得 CA 根憑證，用此來簽署所有需要通訊的憑證。

下載 kubectl-node-shell 進入 docker-desktop 節點並取得 CA 內容：

```
1.  curl -LO https://github.com/kvaps/kubectl-node-shell/raw/master/
    kubectl-node_shell
2.  chmod +x ./kubectl-node_shell
3.  sudo mv ./kubectl-node_shell /usr/local/bin/kubectl-node_shell
4.
5.  kubectl node-shell docker-desktop // 進入 docker-desktop 節點
6.
7.  cat /etc/kubernetes/pki/ca.crt  // 在本地新增一個 ca.crt 檔案將印出的
    內容填入
8.  cat /etc/kubernetes/pki/ca.key // 在本地新增一個 ca.key 檔案將印出的
    內容填入
```

此時我們的目錄底下應該會有以下四個檔案：

```
1.  ls
2.  ------
3.  ca.crt        ca.key        pod-viewer.csr    pod-viewer.key
```

接下來就用拿到的根憑證來產生一個被叢集 CA 簽署過的憑證：

```
1.  openssl x509 -req -in pod-viewer.csr -CA ca.crt -CAkey ca.key
    -CAcreateserial -out pod-viewer.crt -days 365
```

此時我們將會獲得一個 pod-viewer.crt 的憑證，可以使用以下指令來查看詳細
內容：

```
1.  openssl x509 -noout -text -in pod-viewer.crt
2.  ------
3.  Data:
4.          Version: 1 (0x0)
5.          Serial Number:
6.              8f: ac: d9: 57: 79: 80: 11: 8d
7.      Signature Algorithm: sha256WithRSAEncryption
8.          Issuer: CN=kubernetes
9.          Validity
10.             Not Before: Sep 27 10: 24: 39 2022 GMT
11.             Not After: Sep 27 10: 24: 39 2023 GMT
12.         Subject: CN=pod-viewer
13.         Subject Public Key Info:
14.             Public Key Algorithm: rsaEncryption
15.                 RSA Public-Key: (2048 bit)
16.                 Modulus:
17. // ...
```

到這裡我們有了憑證之後就可以開始建立 Context 與 User 了。

使用憑證建立一個 User：

```
1.  kubectl config set-credentials pod-viewer \
2.      --client-certificate=pod-viewer.crt \
3.      --client-key=pod-viewer.key \
4.      --embed-certs=true
5.  -------
6.  User "pod-viewer" set.
```

使用前一章的指令建立一個 Context：

```
1.  kubectl config set-context only-view --cluster=docker-desktop
    --user=pod-viewer
2.  ------
3.  Context "only-view" created.
```

這裡我們建立了一個 Context，指向我們現有的 docker-desktop 叢集，而 user
是還沒被授權的 pod-viewer。

查看 kubeconfig 中的設定一下：

```
1.  kubectl config view
2.  ------
3.
4.  apiVersion: v1
5.  clusters:
6.  - cluster:
7.      certificate-authority-data: DATA+OMITTED
8.      server: https://kubernetes.docker.internal: 6443
9.    name: docker-desktop
10. contexts:
11. - context:
12.     cluster: docker-desktop
13.     user: docker-desktop
14.   name: docker-desktop
15. - context:
16.     cluster: docker-desktop
17.     user: pod-viewer
18.   name: only-view
19. current-context: docker-desktop
20. kind: Config
21. preferences: {}
22. users:
```

```
23. - name: docker-desktop
24.    user:
25.      client-certificate-data: REDACTED
26.      client-key-data: REDACTED
27. - name: pod-viewer
28.    user:
29.      client-certificate-data: REDACTED
30.      client-key-data: REDACTED
```

可以看到我們的 Context 已經設定完畢，但此時我們只完成了 Authentication，並沒有獲得任何權限，可以大膽猜測目前我們可以與叢集溝通，但不能取得任何資源請求回應。

切換到 pod-viewer context 並試圖查看 pod 資訊：

```
1.  kubectl config use-context only-view
2.  -------
3.  Switched to context "only-view
4.
5.  kubectl get pod
6.  -------
7.  Error from server (Forbidden): pods is forbidden: User "pod-viewer"
    cannot list resource "pods" in API group "" in the namespace "default"
```

如預期的 only-view 的 pod-viewer 並沒有擁有任何權限，所以不能通過 Authorization。

2. 使用 RBAC 授權給使用者

在開始操作之前，我們需要對使用 RBAC 進行授權的方式有進一步的了解，它是 Kubernetes v1.8 正式引入的 Authorization 機制，它是一種管制訪問 Kubernetes API 的機制。管理者可以透過 rbac.authorization.k8s.io 這個 API 群組來進行動態的管理設定。主要由 Role、ClusterRole、RoleBinding、ClusterRoleBinding 等資源組成。

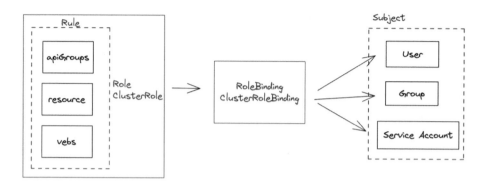

▲ 圖 30-2 RoleBinding

透過適當的角色設定與授權分配，管理者可以決定使用者可以使用哪些功能。
在 RBAC 下的角色會被賦予指定的權限（Permission）並實現最小權限源則，對
比於限制特定權限的方式更為嚴謹。

30.3 Role vs ClusterRole

角色（Role）是一組許可規則的集合，用來定義某個 namespace 內的訪問許
可，而 ClusterRole 則是一個叢集資源。以上兩種不同顆粒細度的資源作用範
圍，可以使我們將權限拆分的更仔細。如果系統管理員沒有 ClusterRole 的權
限，那代表他需要將每個 Namespace 一一綁定到需要叢集等級權限的使用
者，那將是 維運人員的一大夢魘。

就來看看實際例子：

範例檔 **role.yaml**

```
1.   # role.yaml
2.   kind: Role
3.   apiVersion: rbac.authorization.k8s.io/v1
```

```
4.   metadata:
5.     namespace: default # 定義在 default 命名空間
6.     name: pod-viewer        # Role 名稱
7.   rules:
8.   - apiGroups: [""] # "" 預設代表 apiVersion：v1
9.     resources: ["pods"]
10.    verbs: ["get", "watch", "list"]
```

role.yaml 宣告了一個作用在 namespace default 中的 Role 物件，並允許 Role 能夠對 pods 進行限定操作。

構成一個 Rule 需要定義三部分：

● apiGroups：資源所屬的 API 組："" 預設為 core 組資源，例如：extensions、apps、batch 等。

● resources：資源，例如：pods、deployments、services、secrets 等。

● verbs：動作，例如：get、list、watch、create、delete、update 等。

再提供一個 ClusterRole 的例子：

範例檔 **cluster-role.yaml**

```
1.   # cluster-role.yaml
2.   apiVersion: rbac.authorization.k8s.io/v1
3.   kind: ClusterRole
4.   metadata:
5.     name: cluster-pod-viewer
6.   rules:
7.   - apiGroups: [""]
8.     resources: ["pods"]
9.     verbs: ["get", "list", "watch"]
```

很清楚地看出兩者的差異在於 ClusterRole 並不需要指定 Namespace 。

首先需要將 context 切回叢集管理者的 Context，這樣才有權限建立起 Role：

```
1.  # 記得先切換回 docker-desktop
2.  kubectl config use-context docker-desktop
3.  ------
4.  Switched to context "docker-desktop".
5.
6.  kubectl apply -f role.yaml
7.  ------
8.  role.rbac.authorization.k8s.io/pod-viewer created
```

成功建立！但這時我們還未將任何使用者與此 Role 資源綁定，所以接下來就要
設定 RoleBinding 或 ClusterRoleBinding。

▶ 30.4 RoleBinding vs ClusterRoleBinding

前面我們已經擁有了一個帶有授權的 Role，下一步我們需要將此角色綁定到指
定使用者，才能將角色中定義好的授權賦予給一個或一組使用者使用，而以下
的 Subject 代表被綁定的物件。

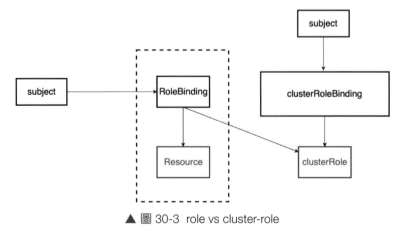

▲ 圖 30-3　role vs cluster-role

被綁定的物件可以是：

- User：對於名稱為 alice@example.com 的用戶。

```
1.  subjects:
2.  - kind: User
3.    name: "alice@example.com"
4.    apiGroup: rbac.authorization.k8s.io
```

- Service Account：對於 kube-system 命名空間中的 default 服務帳戶。

```
1.  subjects:
2.  - kind: ServiceAccount
3.    name: default
4.    namespace: kube-system
```

- Group：在 Kubernetes 中我們可以指定符合特定前綴，將符合條件的使用者劃分為同一組。

```
1.  # 對於"qa" 名稱空間中的所有服務帳戶
2.  subjects:
3.  - kind: Group
4.    name: system: serviceaccounts: qa
5.    apiGroup: rbac.authorization.k8s.io
6.  # 對於所有用戶
7.  subjects:
8.  - kind: Group
9.    name: system: authenticated
10.   apiGroup: rbac.authorization.k8s.io
11. - kind: Group
12.   name: system: unauthenticated
13.   apiGroup: rbac.authorization.k8s.io
```

接下來就讓我們將 RoleBinding 實現出來：

範例檔 **role-binding.yaml**

```
1.  # role-binding.yaml
2.  apiVersion: rbac.authorization.k8s.io/v1
3.  kind: RoleBinding
4.  metadata:
5.    name: pod-viewer-rolebinding
6.    namespace: default #授權的名稱空間為 default
7.  subjects:
8.  - kind: User
9.    name: pod-viewer # 繫結 pod-viewer 使用者
10.   apiGroup: rbac.authorization.k8s.io
11. roleRef:
12.   kind: Role
13.   name: pod-viewer # 繫結 Role
14.   apiGroup: rbac.authorization.k8s.io
```

建立 RoleBinding 資源：

```
1.  kubectl apply -f role-binding.yaml
2.  ------
3.  rolebinding.rbac.authorization.k8s.io/pod-viewer-rolebinding created
```

這時就可以切回我們先前建立好的 only-view 的 Context：

```
1.  kubectl config use-context only-view
2.  ------
3.  Switched to context "only-view".
```

現在就來驗證相關權限是否如預期綁定：

```
1.  # 成功取得 pod 資訊(此時沒有任何 pod 在執行)
2.  kubectl get pod -n default
```

```
3.  ------
4.  No resources found in default namespace.
5.
6.  # 成功收到 forbidden 阻止查看資源
7.  kubectl get pod -n kube-system
8.  Error from server (Forbidden): pods is forbidden: User "pod-viewer"
    cannot list resource "pods" in API group "" in the namespace
    "kube-system"
```

大功告成！我們只要發放這個 Context 給對應的團隊人員，即可以實現方便安全的 RBAC 授權。

筆者碎碎念

到此我們就已經順利的完成了一個完整的 RBAC 流程了，Kubernetes 中在講述權限的篇幅其實非常多，不只是需要對 Kubernetes 資源有一定的了解，更要對認證與授權這些大觀念有深入的概念才可以一窺其妙，加上各種主流或少見的驗證方法，是相對進階的課題了，有興趣的同學真的很推薦把關於這邊的官方文件都啃過一遍，肯定可以跟我一樣讀的頭破血流的。

參考資料

- 管理服務帳號
  https://kubernetes.io/zh-cn/docs/reference/access-authn-authz/service-accounts-admin/

- [Day17] k8s 管理篇（三）：User Management、RBAC、Node Maintenance
  https://ithelp.ithome.com.tw/articles/10223717

- 基於角色的訪問控制（RBAC）
  https://jimmysong.io/kubernetes-handbook/concepts/rbac.html

- RBAC with Kubernetes in Minikube
  https://medium.com/@HoussemDellai/rbac-with-kubernetes-in-minikube-4deed658ea7b

- openSSL 自發憑證
  https://hackmd.io/@yzai/rJXYxFpmq

- 【從題目中學習 k8s】-【Day20】第十二題 - RBAC
  https://ithelp.ithome.com.tw/articles/10244300

- Day 19 - 老闆！我可以做什麼：RBAC
  https://ithelp.ithome.com.tw/articles/10195944

- 使用 RBAC 鑑權
  https://kubernetes.io/zh-cn/docs/reference/access-authn-authz/rbac/#referring-to-subjects

- k8s 基於 RBAC 的認證、授權介紹和實踐
  https://iter01.com/657295.html

博碩文化

博碩文化